D0121840

SALT WATER INTRUSION

Status and Potential in the Contiguous United States

by

S. F. Atkinson
G. D. Miller
D. S. Curry
S. B. Lee

Environmental and Ground Water Institute
University of Oklahoma
Norman, Oklahoma

Library of Congress Cataloging-in-Publication Data

Salt Water Intrusion

    Bibliography: p.
    Includes index.
    "List of counties by state and aggregated
subareas"—P.
    1. Salt water encroachment—United States.
I. Atkinson, S. F. (Sam F.)
GB1197.82.S25   1986   363.7′394   86-7344
ISBN 0-87371-054-1

COPYRIGHT © 1986 by LEWIS PUBLISHERS, INC.
ALL RIGHTS RESERVED

Neither this book nor any part may be reproduced or transmitted in
any form or by any means, electronic or mechanical, including
photocopying, microfilming, and recording, or by any information
storage and retrieval system, without permission in writing from the
publisher.

LEWIS PUBLISHERS, INC.
121 South Main Street, P.O. Drawer 519, Chelsea, Michigan 48118

PRINTED IN THE UNITED STATES OF AMERICA

SAMUEL F. ATKINSON is an Environmental Engineering Scientist with the Environmental and Ground Water Institute at the University of Oklahoma in Norman and is currently under a full-time contract to the U.S. Army Corps of Engineers. He received his Ph.D. in Engineering/Environmental Science in 1985 from the University of Oklahoma, with a minor in Regional and City Planning. His M.S. in Environmental Science was received in 1982 from the University of Oklahoma, and his B.S. in Biology in 1979 from Oklahoma State University in Stillwater. He was initiated into Chi Epsilon (the National Civil Engineering Honor Society) in 1983. Before coming to the University of Oklahoma, he was a laboratory research technician for a major oil company, and previous to that he was a teaching assistant at Oklahoma State University.

Dr. Atkinson has studied and conducted research on numerous environmental impact assessment topics. His work with the U.S. Army Corps of Engineers involves water resources planning for the Fort Worth, Texas District using advanced remote sensing techniques. Another aspect of his research involves ground water quality management and pollution control. His dissertation involved a simplified technique for biological impact assessment using remotely sensed imagery. He coauthored *Impact of Growth*, also by Lewis Publishers, and has written and presented several papers on environmental impact assessment, remote sensing, and ground water quality.

Gary D. Miller recently became Assistant Director and Research Program Coordinator of the Hazardous Waste Research and Information Center within the Illinois State Water Survey. Previously, he was Assistant Professor of Civil Engineering and Environmental Science at the University of Oklahoma. He taught graduate level courses in Ground Water Quality Management and was principal investigator for several research projects related to the chemical and biological fate of contaminants in ground water. Dr. Miller is coauthor of three books and more than twenty-five other publications. His research interests include environmental chemistry and biology, hazardous waste treatment and disposal, resource management, and risk assessment.

Debra Songer Curry is a Research Assistant for the Environmental and Ground Water Institute. She is currently pursuing a Master's Degree of Environmental Science in Ground Water Quality Management at the University of Oklahoma. She received her Bachelor of Science in Geology at the University of Oklahoma and worked two years as a coal geologist.

Sa Ba Lee is currently pursuing his Ph.D. in Civil Engineering and Environmental Science at the University of Oklahoma with a minor in Regional and City Planning. He received his Bachelor of Technology Degree in 1977 from the Indian Institute of Technology, Madras, India, and a M.E. in 1979 from Asian Institute of Technology, Bangkok, Thailand. He has worked as a civil/environmental engineer for a major engineering consulting firm in Malaysia. He is presently a Research Assistant working with the Environmental and Ground Water Institute. He has participated in numerous research projects undertaken by the Institute.

PREFACE

Salt water can move into fresh ground water supplies as a result of overpumping in coastal and tidal areas, overpumping in noncoastal areas which overlay saline water, movement of saline water through leaky well casings, and via natural processes such as tidal variations or drought. Salt water intrusion into fresh ground water can limit the uses of ground water for agricultural, industrial, or domestic purposes. Due to the increasing demands for usage of ground water resources in the contiguous United States, it is important to evaluate and control salt water intrusion into these valuable resources.

This book provides a summary of the status and potential for salt water intrusion into ground water in the contiguous United States. While the focus is on resultant limitations in the agricultural usage of ground water, the book is not limited to this singular limitation in resource usage. The book is organized into five chapters and two appendices. Key chapters or chapter sections are included on the effects of salinity, sources of salt water intrusion, physical control methods, results of a survey of the national status of salt water intrusion, and a systematic methodology for considering current and future salt water intrusion problems by water resources aggregated subareas. The key appendix contains an annotated bibliography of 125 pertinent references identified from computer-based searches of the literature.

The authors wish to express their appreciation to a number of individuals involved in a variety of ways in the development of this book. First, the support of the U.S. Soil Conservation Service (SCS) and the assistance of Clive Walker from the Washington headquarters office are acknowledged. Dennis Erinakes of the SCS Technical Center in Fort Worth, Texas, provided valuable guidance and suggestions via a series of meetings with the authors.

Several research staff members of the Environmental and Ground Water Institute (EGWI) at the University of Oklahoma participated in information gathering and/or discussions related to the development and organization of this book. First, Dr. Larry Canter, Sun Company Professor of Ground Water Hydrology and Director, EGWI, prepared Appendix A containing the annotated bibliography of 125 references. Providing assistance in a more general sense were Deborah Fairchild, Environmental Scientist; and Margaret O'Hare and Paris Hajali, both graduate research assistants. The able typing assistance of Mrs. Leslie Rard and Mrs. Wilma Clark is gratefully acknowledged. Finally, and most important, the authors are indebted to Mrs. Mittie Durham of EGWI for her typing skills and devotion to the preparation of this manuscript.

The authors also wish to express their appreciation to the University of Oklahoma College of Engineering for its basic support of faculty and research staff writing endeavors, and to their families for their understanding and patience.

Sam Atkinson
Environmental Scientist
Environmental and Ground
   Water Institute
University of Oklahoma
Norman, Oklahoma

Gary Miller
Assistant Professor of
   Environmental Science
School of Civil Engineering
   and Environmental Science
University of Oklahoma
Norman, Oklahoma

Debra S. Curry
Graduate Research Assistant
Environmental and Ground
   Water Institute
University of Oklahoma
Norman, Oklahoma

Sa Ba Lee
Graduate Research Assistant
Environmental and Ground
   Water Institute
University of Oklahoma
Norman, Oklahoma

To Pye TeSelle
To Gina, Gwenna, and Gregory
To Paul and Savilla
To my mother and father

CONTENTS

FIGURES

TABLES

CHAPTER 1

INTRODUCTION

The control of salt-water intrusion into fresh ground water is a basic concern in many parts of the United States relative to long-term dependable water supplies. Causes of salt-water intrusion include overpumping in coastal and tidal areas, overpumping of fresh water supplies which overlay saline water, leaky well casings which can promote the mixing of water from a saline aquifer with fresh water in an overlying or underlying aquifer, and natural sources or processes, such as tidal variations or even drought.

The effects of salt-water intrusion related to human health and welfare include decreased water quality and desirability for drinking water. In agricultural areas increasing salinity in irrigation water results in declining crop productivity and may necessitate switching to salt-tolerant crops. Wildlife can also be affected by high saline runoff or from increased salinity of springs used for watering. Highly saline waters can also be undesirable for many industrial purposes.

SALINITY CONCERNS

Salinity can be technically defined as the total mass in grams of all dissolved substances per kilogram of water, with all carbonate converted to oxide, all bromide and iodine replaced by chlorine, and all organic matter oxidized at 480ºC. Salinity is related to the mineral content or the dissolved solids contained in water. The U.S. Public Health Service in 1962 established a drinking water standard for total dissolved solids of 500 mg/l. The revised drinking water standards issued by EPA include a secondary standard of 500 mg/l to protect welfare but no primary total dissolved solids standard to protect health. Waters have been classified according to their dissolved solids content for various uses. One classification of water based on salinity is as follows (Summers and Schwab, 1970):

NAME	SALINITY (IN PARTS PER THOUSAND)
Fresh Water	0.5
Brackish Water	
Oligohaline	0.5 - 3.0
Mesohaline	
Meiomesohaline	3.0 - 8.0
Pleiomesohaline	8.0 - 16.5
Polyhaline	16.5 - 30.0
Marine Water	30.0

For irrigated agriculture water can be classified according to salt content:

CLASSIFICATION	SALT CONCENTRATION (IN MG/L)	
Excellent	175	
Good	175	- 525
Permissible	525	- 1,400
Doubtful	1,400	- 2,100
Injurious	2,100	

Plants vary in their tolerance to salinity and some states recognize this in their water quality criteria. For example, in Minnesota a limit of 700 mg/l has been established for waters used for rice production (Class A) and 1,000 mg/l for other agricultural waters.

Increasing salinity in fresh water aquifers can occur, and has occurred, in numerous locations throughout the United States. In fact, in almost every state there is at least one area of salt-water intrusion concern, or area where natural salinity is a potential problem. From an historical perspective, salt water in coastal areas was identified as the initial focus of increasing salinity in ground water supplies. In recent years it has been well documented that there are other potential sources of salt water which can move into fresh ground water aquifers. Examples of this include the movement of salt water to fresh water zones from leaking wells, transport of salts from surface waste disposal sites into the fresh water zone, and discharges from saline aquifers as a result of inappropriate spacing of injection wells relative to existing wells in a given geographical area.

The primary concern relating to increasing salinity in ground water is associated with potential limitations on ground water usage. This is important throughout the United States relative to ground water usage for agricultural purposes.

OBJECTIVE OF THIS BOOK

This book addresses the issue of salt water intrusion as it relates to future uses of ground water resources for agricultural and other purposes. Therefore, emphasis is given to the extent of salt water intrusion resulting from coastal intrusion, leaking wells, upconing of salts into aquifers due to over-pumping, surface sources of salt leaking into aquifers, penetration of confined salty aquifers by drilling, and discharges from saline aquifers to the surface. This survey of salt water intrusion can aid in identifying limitations of the uses of ground water as a result of increasing ground water salinity.

CONTENTS OF THIS BOOK

This book is organized into five chapters and two appendices. Following this introductory chapter, Chapter 2 provides a detailed description of effects of salinity and sources of salt water intrusion. Salinity evaluation via monitoring and modeling, physical control measures, and institutional and legal aspects of management are summarized in Chapter 3. Chapter 4 summarizes the results of a national salinity survey. A systematic methodology for considering current and future salt water intrusion problems by water resources aggregated sub-areas is described in Chapter 5.

Appendix A contains an annotated bibliography of 125 references identified from computer-based searches of the literature. Each included abstract is categorized into one of the following 12 categories: bibliographic materials; overview information on salt water intrusion; coastal salt water intrusion; noncoastal intrusion of salt water; oil-gas field activities; irrigation effects; fresh water (injection and surface water pressure-head) as a barrier to seawater; institutional measures; modeling: coastal seawater intrusion-- physical; modeling: coastal seawater intrusion--numerical; modeling: noncoastal salt water intrusion--physical; and modeling: noncoastal salt water intrusion--numerical. Finally, Appendix B contains a listing of all counties in the continental United States organized by state and aggregated subareas.

SELECTED REFERENCE

Summers, W.K., and Schwab, G.E., "A Survey of Saline Ground Water as a Mineral Resource", in Proceedings of the Symposium on Groundwater Salinity, R. B. Mattox, ed., 46th Annual Meeting of the Southwestern and Rocky Mountain Division of the American Association for the Advancement of Science, 1970, Las Vegas, Nevada, pp. 31-45.

CHAPTER 2

DESCRIPTION OF EFFECTS AND SOURCES
OF SALT WATER INTRUSION

The deleterious impact of high salinity ground water on municipal, industrial, and agricultural water users is felt in a number of ways, including high soap consumption; formation of scale in heating vessels; corrosive attack on distribution pipelines, plumbing systems and appliances; and added water treatment and conditioning costs. A number of estimates have been made as to the added costs to municipal, industrial, and agricultural users due to increases in salinity. It is difficult to place precise dollar values on the impacts; however, one study (Holburt, 1972) estimated that salinity alone would impact the Lower Colorado River Basin in the order of $45 to $60 million per year by the turn of the century.

Optimum or tolerable salinity concentrations in ground water are different for various uses. Waters with less than about 500 mg/l total dissolved solids (TDS) are generally considered suitable for domestic purposes, while waters with concentrations greater than about 5000 mg/l TDS generally are unsuitable for irrigation purposes. Livestock water consumption requirements can be met with water from less than 3,000 mg/l TDS for poultry to as much as 12,000 mg/l TDS for sheep (U.S. Environmental Protection Agency, 1973). Summers and Schwab (1970) list the following limiting concentrations of dissolved solids:

Use	Limit in mg/l
Boiler-feed water	50-3,000 (depending on pressure)
Brewing, light beer	500
Brewing, dark beer	1,000
Brewing and distilling, general	500-1,500

In most cases, the salinity of ground water is a prime determinant in its use. The remainder of this chapter expands the discussion of the problem of salt water intrusion. This description of the problem begins with a discussion of the effects of salinity on plants and crops, briefly describes various other effects, and finally provides information on the sources of salt water contamination of ground water.

EFFECTS OF SALINITY

Salinity effects can be considered in terms of plant response, crop tolerance, and other effects.

Plant Response

As the salinity of a plant's ground water supply increases, plant growth usually is reduced. There have been many attempts to explain how the growth reduction is caused by a salinity increase. Most of those attempts have invoked the concept of physiological drought in some form or other. However, in view of recent results in physiological studies, such a view no longer holds. On the other hand, concepts which rely heavily on specific ion effects rather than physiological drought do not explain the many observations that plant growth is more directly related to the salt concentration or osmotic pressure of the ground water. That is, even though chloride salts are more toxic than sulfate salts under some conditions, there are occasions when iso-osmotic concentrations of different salts give similar responses.

In response to the differing views on how salinity affects plant growth, O'Leary (1970) attempted to present a unified theory. The theory is based on an equation which relates water movement from the outside of a plant's root to the inside of the root's xylem cells. The equation relates water flow to the permeability of the plant, the pressure inside and outside of the plant, the osmotic pressure of the water inside and outside of the root cells, and how effective the plant is at excluding solutes. The equation is (O'Leary, 1970):

$$J_v = L_p \left[(P_o - P_i) - \sigma(\pi_o - \pi_i) \right]$$

where J_v = volume flow ($cm^3 cm^{-2} sec^{-1}$)

L_p = permeability coefficient of hydraulic conductivity ($cm\ sec^{-1} bar^{-1}$)

P = pressure in excess of atmospheric pressure (bars)

π = osmotic pressure (bars)

σ = reflection coefficient which expresses the effectiveness of the roots in excluding solutes.

The subscripts, o and i, indicate the outside and inside of the root, respectively.

Under most conditions, in a transpiring plant, P_o is zero, P_i has a negative value (i.e., the solution in the xylem is under tension), and π_o and π_i are slightly greater than zero. The reflection coefficient at the root surface can be considered as unity. Thus, the above equation can be reduced to:

$$J_v = L_p(\Delta P \cdot \Delta \pi)$$

or $J_v = L_p(\Delta \Psi)$

or   Flow = Permeability x Driving Gradient

As ground water salinity increases, and π_o becomes higher, the driving gradient for water movement into the root becomes less. This is the basis for early ideas that the cause of growth inhibition was physiological drought. If the salinity increase is a sudden one, such a situation does, in fact, occur. However, under most conditions, even for halophytes, the salinity increase is gradual. Under this condition, most plants have been found to increase the internal osmotic pressure so the apparent driving gradient is thereby maintained. This is true for both root cells and leaf cells. That is, as π_o increases, so does π_i to a degree. Nevertheless, the symptoms of plants subjected to increased ground water salinity often have been observed to resemble the symptoms of drought. From the above expressions it can be seen that decreased water flow could result from a decrease in root permeability, even if the driving gradient is maintained.

With this consideration in mind, O'Leary (1970) looked at root permeability under saline conditions and found that it is reduced considerably by small increases in salinity. This reduction in root permeability was sufficient to reduce the relative water content of leaves, and this reduction was enough to apparently reduce the stomatal aperature. The reduction in stomatal aperature results in less transpirational loss of water, and the plant re-establishes a water balance-equilibrium.

O'Leary (1970), knowing that there is some minimal turgor pressure necessary for leaf expansion (i.e., growth), claimed that as long as the leaves are suffering from water deficit, leaf expansion will be suppressed. Furthermore, if the evaporative demand of the atmosphere is high enough, partial stomatal closure may not even prevent a water deficit from occurring, and the plant will give the typical appearance of a plant suffering from drought. However, if the evaporative demand of the air is low, the plant should be able to maintain a favorable water balance in spite of low rates of water absorption because of the reduced transpirational loss. O'Leary (1970) tested his hypothesis by growing plants in closed microcosms where the atmospheric humidity was maintained at an extremely high level. The expected results occurred.

High humidity did not eliminate salinity-induced growth inhibition completely, however, thus indicating some direct effect of salinity on growth. To identify the direct effect of salinity on plant growth, O'Leary (1970) considered plants which do not suffer from water stress. These types of plants include halophytes, and the typical response of halophytes to increased salinity is increased succulence. O'Leary (1970) then addressed the question of why halophytes seem to have more than enough water. In a saline environment, a halophyte will accumulate some of the salts, which will create a higher osmotic pressure outside of the plant than inside. This condition will force an increased influx of water. The leaves did not expand; they became more succulent. If leaf expansion is a turgor pressure driven process, then there are basically two ways that cell expansion could be suppressed (O'Leary, 1970):

(1)   not enough water absorption by the cells to develop sufficient turgor pressure; or

(2)   decreased extensibility of the cell walls so that even if enough water is absorbed, the cell walls resist stretching.

O'Leary (1970) found increasing evidence that shows that the symptoms of water and salt stress often resemble the symptoms of senescence (old age) in leaves. As such, it was suggested that water and salt stress might accelerate the onset of senescence. Further, as cells mature, cell wall extensibility decreases due to secondary wall thickening. Therefore, the higher turgor pressure in leaves of plants subjected to increasing salinity may be the result of an accelerated rate of maturation and even senescence in those cells.

The presence of cytokinin-like substances has been detected in roots and in the bleeding stream from truncated root systems, and they appear to be continuously synthesized in roots. Since cytokinins seem to control activities in the leaves, particularly those associated with development of senescence, it seems likely that a normal feature of a plant's physiology involves synthesis of cytokinins or similar substances in the roots and subsequent transport to the leaves. Furthermore, any perturbation of this activity would be expected to result in a disruption of hormone balance in the leaves and possibly even an acceleration of the onset of senescence.

O'Leary (1970) found that less cytokinin was recovered from root systems subjected to water stress, and that the amount of cytokinins reaching the shoots under normal conditions drops off significantly when leaf senescence begins. Therefore, he hypothesized that a significant effect of increasing salinity in the root environment is a decreased delivery of cytokinins to the leaves.

The generalized scheme is shown in Figure 1 (O'Leary, 1970). The indication is that ground water salinity would have two separate but related effects on plant growth. One effect is that as salinity is increased plants that are inefficient at accumulating salts (non-halophytes, for example) exclude more and more salt, and the osmotic pressure within the plant becomes higher than the osmotic pressure outside of the plant. The result is that water tends to flow out of the root system causing a physiological drought. Secondly, as the plant becomes more stressed, cytokinin production is reduced, and leaves begin senescence, thus further stressing the plant.

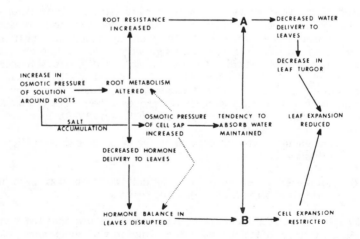

Figure 1: Generalized Scheme of How Ground Water Salinity Affects Plant Growth (O'Leary, 1970)

Crop Tolerance

Most higher plants can be divided into two distinct groups: monocotyledons and dicotyledons. A cotyledon is a fleshy leaf-type structure which is apparent when a seedling emerges from the ground. The cotyledon, commonly called "seed leaves", is a source of stored food which the seedling utilizes during early growth. By the time a plant is capable of converting enough sunlight to satisfy growth requirements, the supply of food in the cotyledons is exhausted and they eventually fall off. The distinction between mono- and dicotyledons, as the name implies, is whether the seedling has one or two cotyledons.

Plants may also be divided into two general groups, halophytes and glycophytes. The distinction between the groups is based on a plant's tolerance to salt; halophytes are salt-tolerant; glycophytes are salt-sensitive. The classification between the two groups is, however, not absolute because plants can range from highly salt-tolerant to highly salt-sensitive.

Lauchi and Epstein (1984) have pointed out that most monocotyledonous halophytes show reduced growth rates at salinity levels exceeding about 10,000 mg/l solium chloride. Dicotyledonous halophytes, however, exhibit growth until salinity levels are increased to about 15,000 mg/l sodium chloride. Most crop species, however, belong to the glycophytic group. Lauchi and Epstein (1984) also point out that most glycophytes respond negatively to salt concentrations above 6000 mg/l. Some species (carrots, for example) are sensitive to salt concentrations as low as 640 mg/l.

The U.S. Department of Agriculture Salinity Laboratory in Riverside, California, has been conducting laboratory and field tests on the salt tolerance of crops since its establishment in 1937. Salt tolerance tests are usually conducted in small experimental plots, where commercial practices are followed as closely as possible, with adequate moisture and fertility. To ensure an acceptable stand, researchers plant seed in a nonsaline seedbed and impose salinity by adding calcium and sodium chloride salts to the irrigation water after the seedlings have emerged. They test several salinity levels to determine both the threshold level that begins to decrease yield and the rate of yield reduction caused by higher levels. Generally, the higher the threshold level, the less yield is decreased as salinity increases.

The U.S. Department of Agriculture Salinity Laboratory has also found that a variety of environmental factors (including plant, soil, and weather conditions) affect crop growth. Thus, salt tolerance is measured as a function of growth, or yield. Therefore, salinity impacts reduce the yield of a crop, and the impact is expressed as the percent reduction of yield experienced by crops in saline versus nonsaline conditions (where environmental conditions are similar). Although yields vary from location to location and year to year, the relative yield reductions caused by salinity remain reasonably consistent.

Soil salinity in the plant root zone is measured by the U.S. Department of Agriculture Salinity Laboratory as electrical conductivity, which is directly proportional to the salt concentration in the soil water. Two commonly used methods provide reliable measurements: one involves sampling the soil within the root zone, preparing a saturated extract, and measuring the electrical conductivity of the extracted soil water; the other uses a recently developed instrument that, when inserted into the soil, directly measures the electrical conductivity of the soil water. Since soil salinity usually increases with depth, measurements are taken at several depths within the root zone, and the values averaged (Maas, 1984). When the electrical conductivity of a saturated soil extract is measured, the units are decisiemens per meter (dS/m), where 1dS/m is approximately equivalent to 640 mg/l salt. Table 1 shows the salt tolerance of agricultural crops as determined by the U.S. Department of Agriculture Salinity Laboratory (Maas, 1984).

Table 1: Salt Tolerance in Agricultural Crops (Maas, 1984)

Salt Tolerance of Agricultural Crops

Crop	Maximum soil salinity without yield loss (threshold)*	% Decrease in yield at soil salinities above the threshold
Sensitive crops	dS/m	% per dS/m
Bean	1.0	10
Carrot	1.0	14
Strawberry	1.0	33
Onion	1.2	16
Almond#	1.5	19
Blackberry	1.5	22
Boysenberry	1.5	22
Plum, prune#	1.5	18
Apricot#	1.6	24
Orange	1.7	16
Peach	1.7	21
Grapefruit#	1.8	16
Moderately senstitive crops		
Turnip	0.9	9.0
Radish	1.2	13
Lettuce	1.3	13
Clover, berseem	1.5	5.7
Clover, strawberry	1.5	12
Clover, red	1.5	12
Clover, alsike	1.5	12
Clover, ladino	1.5	12
Foxtail, meadow	1.5	9.6
Grape#	1.5	9.6
Orchardgrass	1.5	6.2
Pepper	1.5	14
Sweet potato	1.5	11
Broadbean	1.6	9.6
Corn	1.7	12
Flax	1.7	12
Potato	1.7	12
Sugarcane	1.7	5.9
Cabbage	1.8	9.7

Table 1: (Continued)

Salt Tolerance of Agricultural Crops

Crop	Maximum soil salinity without yield loss (threshold)*	% Decrease in yield at soil salinities above the threshold
Celery	1.8	6.2
Corn (forage)	1.8	7.4
Alfalfa	2.0	7.3
Spinach	2.0	7.6
Trefoil, big	2.3	19
Cowpea (forage)	2.5	11
Cucumber	2.5	13
Tomato	2.5	9.9
Broccoli	2.8	9.2
Velch, common	3.0	11
Rice, paddy@	3.0+	12+
Squash, scallop	3.2	16
Moderately tolerant crops		
Wildrye, beardless	2.7	6.0
Sundangrass	2.8	4.3
Wheatgrass, std. crested	3.5	4.0
Fescue, tall	3.9	5.3
Beet, red@	4.0	9.0
Hardingrass	4.6	7.6
Squash, zucchini	4.7	9.4
Cowpea	4.9	12
Soybean	5.0	20
Trefoil, birdsfoot	5.0	10
Ryegrass, perennial	5.6	7.6
Wheat, durum	5.7	5.4
Barley (forage)@	6.0	7.1
Wheat@	6.0	7.1
Sorghum	6.8	16
Tolerant crops		
Date palm	4.0	3.6
Bermudagrass	6.9	6.4
Sugarbeet@	7.0	5.9
Wheatgrass, fairway crested	7.5	6.9
Wheatgrass, tall	7.5	4.2
Cotton	7.7	5.2
Barley@	8.0	5.0

Table 1: (Continued)

NOTE: These data serve only as guidelines to relative tolerances among crops.   Absolute tolerances vary, depending on climate, soil conditions, and cultural practices.   **SOURCE:**   Maas, E.V., 1984, Salt Tolerance of Plants, in Handbook of Plant Science in Agriculture, (ed.) B. R. Christie, CRC Press Inc., Boca Raton, Florida (in press). *Soil salinity expressed as electrical conductivity of saturated soil extracts.   1 decisiemens per meter (dS/m) = 1 millimho per centimeter (mmho/cm).   1 dS/m = approximately 640 mg/L salt.   #Tolerance is based on growth rather than yield.   @Less tolerant during emergence and seedling stage.   +Values for paddy rice refer to the electrical conductivity of the soil water during the flooded growing conditions.

The following discussion of the effects of salinity on crops is based on the work of E.V. Maas, a plant physiologist at the U.S. Department of Agriculture Salinity Laboratory (Maas, 1984).   In the agricultural environment, salinity levels often change throughout the season.   Although most crops become more tolerant at later stages of growth, there are some exceptions.   For example, salt seems to affect rice during pollination and may decrease seed set and grain yield.   Plants are generally more sensitive during the seedling and early vegetative stages of growth.   The U.S. Department of Agriculture Salinity Laboratory is conducting experiments on the relative sensitivity of several crops at different stages of development.   In sweet corn, for example, it was found that, although seedling growth was reduced by salinity, the salt level of the irrigation water could be increased up to about 9 dS/m (5,800 mg/l) during the tasseling and grain-filling states without affecting yield.

Because of the greater sensitivity at seedling emergence, it is imperative to keep salinity levels in the seed bed as low as possible at planting time.   If salinity levels reduce plant stand, potential yields may be decreased far more than predicted by salt tolerance data.

Salt tolerance information usually applies to crops irrigated by surface methods, such as furrow or basin-type flooding.   Sprinkler-irrigated crops are subject to damage by both soil salinity and salt spray to the foliage.   Salts may be directly absorbed by the leaves, resulting in injury and loss of leaves.   In crops that normally restrict salt movement from the roots to the leaves, foliar salt absorption can cause serious problems not normally encountered with surface irrigation systems.   For example, water with about 4.5 dS/m (2,900 mg/l) salts reduced the yield of peppers by over 50 percent when sprinkler-applied, but only 16 percent when applied to the soil surface.

Unfortunately, no information is available to predict yield losses as related to salinity levels in sprinkler irrigation water.   Greenhouse experiments conducted at the U.S. Department of Agriculture Salinity Laboratory to test the susceptibility of various crops to foliar injury from saline sprinkling waters indicate that tolerance is related more to the amount of salt absorbed by the leaves than to their tolerance to soil salinity.   The degree of injury depends on weather conditions and water stress.   For instance, leaves may contain excessive levels of salt for several weeks without any visible injury symptoms and then become severely burned when the weather becomes hot and dry.

It is reasonable to assume that saline irrigation water will reduce yields of sprinkled crops at least as much as those of surface-irrigated crops. Additional reductions in yield could be expected for crops susceptible to sprinkler-induced foliar injury.

Unlike most annual crops, tree fruit crops are sensitive to specific salt constituents and often develop leaf burn symptoms from toxic levels of sodium or chloride. Different varieties and rootstocks accumulate these ions at different rates, and so each one must be evaluated individually. Because it is difficult to obtain such data on producing trees, the figures are usually based on vegetative growth of young trees rather than on yield; consequently, they provide only general guidelines (Maas, 1984).

Other Effects

Saline ground water can have effects other than impacts on plants and crop production. One of these effects is the influence of salinity on soil erosion. Jenkins and Moore (1984) describe the erosion of soils as a surface phenomenon involving the removal and entrainment of particles due to fluid flow-induced shearing stresses. They point out, however, that the nature of these forces resisting these applied stresses is quite different for two general categories of soils termed cohesionless and cohesive. "Cohesionless" soils are coarse-textured materials comprised of sands and gravels. On the other hand, "cohesive" soils are fine-textured materials predominantly comprised of silts and clays with various amounts of fine sand. Due to the relatively large size and inert surface potential of cohesionless soil particles, erosion is predominantly resisted by gravitational forces. These forces also contribute to cohesive soil resistance but to a much lesser extent. Instead, the erosion resistant forces of cohesive soils are attributed to the internal cohesion developed by net attraction-repulsion potentials between electrically-charged clay particle surfaces and edges.

Unlike the erosion mechanism of cohesionless soils controlled purely by physical parameters of the system, the cohesive soil eroding mechanism is complex, involving physico-chemical interactions between the soil matrix and environment. Since few attempts had been made to quantify and predict cohesive soil erosion, Jenkins and Moore (1984) conducted a research project to identify the factors affecting cohesive soil stability. They devised a laboratory procedure which utilized a small-scale, straight-channel, recirculating flume to study the effects of salinity on cohesive soil erosion. The fabricated flume apparatus satisfactorily measured the gross effect of increasing total salt concentration on cohesive soil erodibility at a high degree of statistical significance (Jenkins and Moore, 1984).

Jenkins and Moore (1984) found that cohesive soils demonstrated increased stability and resistance to erosion as the total salt concentration in the eroding water increased. The stabilizing effect of increasing salt was attributed to suppression of interparticle repulsive forces and the existence of constant attractive forces (electrostatic double-layer forces) which dictate flocculation/agglomeration characteristics of fine-grained soils. However, the effect of specific chemical interactions on cohesive soil erodibility remains speculative. The data collected do not support or reject the hypothesis that specific chemical interactions may constitute an additional mechanism affecting erodibility.

Implications for salt water intrusion can be seen in Jenkins and Moore's research (Jenkins and Moore, 1984). They point out that cohesive soils are typical of marshland areas. These marshland environments are frequently altered by an influx of fresh water (for example, seasonal variations in fresh river flow into estuaries and storm-water runoff collected from development areas and disposed in marine lagoons). An increased erodibility may have serious effects on the stability of unlined ditches, drainage canals and embankments of navigable waterways. Since the designed usage of these structures would be hindered by siltation from increased erosion, maintenance costs would be substantial to clean and dredge canals and channels. Therefore, in marshland areas where salt water intrusion control plans are considered, a decrease in salinity may result in increased erosion. A careful assessment would be required to determine if the benefits from salt water intrusion control would outweigh the impacts of increased erosion.

Another potential effect of saline ground water relates to its use as drinking water and the resultant chronic health impacts. Early studies of the relationship between the mineral content of water and apoplexy in Japan, and cardiovascular disease in the United States, raised the possibility that a so-called water factor was involved (Pye, Patrick, and Quarles, 1983). Hubert and Canter (1980) indicated, however, that the chronic effects of mineralized ground water are probably due to life style rather than environmental factors. Although many of the trace elements associated with saline ground water are necessary for good health, high concentrations may be harmful. Excessive concentrations of certain minerals have been implicated in the various chronic forms of cardiovascular disease and cancer, but the studies to date are not conclusive.

SOURCES OF SALT-WATER INTRUSION

This section provides information on the most common sources of salt-water intrusion into fresh water aquifers (Task Committee on Saltwater Intrusion, 1969). There are various sources of salinity which can potentially adversely affect ground water quality which are not discussed in this section. Since this effort was directed at nonsurface sources of salt-water intrusion (sea water intrusion, leaking wells, upconing, and natural salinity), only minimal information was collected on surface salt sources. It is recognized, however, that some surface sources can impact fresh ground water as adversely as subsurface sources. For example, oil field brine pits can be a significant source of salinity as waste brine leaches from the surface to underlying fresh ground water. Road de-icing salts which are stored in large piles can also generate saline leachates.

In coastal areas, sea water can intrude not only from the subsurface, but from the surface. As the high tide moves into an area, sea water moves into canals and shipping channels and travels inland. This upstream migration of sea water alters the salinity of aquatic habitats and may render fish and wildlife habitats unsuitable for native populations. This phenomenon has been observed during times of drought also. Since less fresh ground and surface water is discharged to the sea, sea water can migrate further inland than during wet periods. Fish and wildlife populations in southern Florida are faced with this situation (Erinakes, 1985).

In the western United States, irrigation is one of the largest sources of saline ground water. The problems center on the accumulation of soil salts on croplands as a result of irrigation activities. Water which is used for irrigation will enter the soil, depositing salts as the water is utilized by the crops. The process of evapotranspiration in most crops results in absorbing water from the soil but filtering out high salt concentrations. Finally, soils become so saline that crops will no longer grow in the area. However, the salt in the soil can be leached into underlying fresh ground water, thus further complicating the situation.

The remainder of this section will summarize subsurface sources of salt-water intrusion.

Sea Water

In many areas of the conterminous United States, saline and fresh subsurface waters occur in close proximity. This proximity implies an inherent danger of saline contamination of fresh water resources. In many localities such contamination is a reality. When fresh ground water is pumped from aquifers that are in hydraulic connection with the sea, the gradients that are set up may induce a flow of salt water from the sea toward the well. This migration of salt water into freshwater aquifers under the influence of ground water development is known as sea water intrusion.

The earliest attempts to define the processes involved in sea water intrusion were conducted by Ghyben and Herzberg, two European scientists, in the late 1800's. They found (independently) that the depth to salt water in coastal aquifers was equal to 40 times the elevation of the fresh water above sea level. Ghyben (1888) described the fresh and salt waters as two immiscible fluids separated by a sharp interface. However, due to convectional dispersion and molecular diffusion, the waters mix in a transitional zone of varying degrees of salinity.

Figure 2 illustrates the natural conditions of fresh water/sea water zones (Todd, 1974). Due to the higher density of sea water, it tends to force its way underneath fresh water. However, since the piezometric head of the fresh water is higher than that of sea water, the fresh water is continually discharged to the sea. The fresh water discharge offsets the inland movement of sea water, and an equilibrium is established. Sea water intrusion occurs when fresh water is withdrawn. The withdrawal changes the equilibrium between fresh and sea water since the fresh water piezometric head is decreased. The decrease in the fresh water piezometric head allows sea water to move further inland, thus intruding where formerly fresh water existed.

Figures 3 and 4 indicate the situation which occurs when fresh water is withdrawn from both an unconfined and a confined fresh water aquifer (Todd, 1974). Both figures illustrate the natural equilibrium between fresh and sea water and what happens as the piezometric head of the fresh water is decreased (the piezometric head is equivalent to the water table in an unconfined aquifer).

The piezometric head of fresh water can also be reduced by decreasing the recharge of the aquifer. For example, as a coastal area becomes more and more urbanized, surface runoff increases. The increase in surface runoff results in less infiltration, thus reducing recharge. Ultimately, there is less ground water flow to the sea which allows sea-water intrusion to occur.

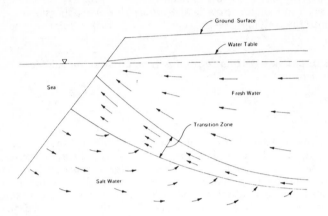

Figure 2:   Relationship Between Sea Water and Fresh Ground Water Under Natural Conditions

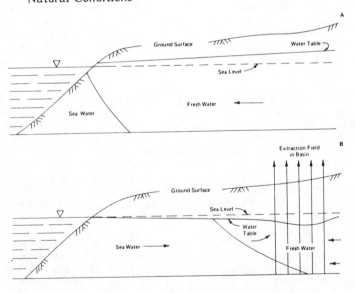

Figure 3:   Hydrological Conditions in an Unconfined Coastal Ground Water Basin:

A--Not Subject to Sea-water Intrusion
B--Subject to Sea-water Intrusion

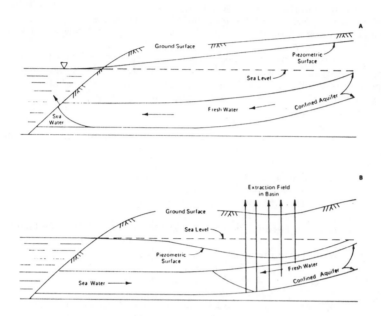

Figure 4:   Hydrological Conditions in a Confined Aquifer in a Coastal Area:

A--Not Subject to Sea-water Intrusion
B--Subject to Sea-water Intrusion

Many coastal areas in the United States have experienced sea-water intrusion due to both increased ground water withdrawal and increased urbanization. Long Island and mainland New York provide one example (Ballentine, Reznek, and Hall, 1972). Major intrusions have occurred as a result of ground water withdrawal and the decrease in aquifer recharge caused by improved drainage and sewer systems. By the mid 1930's, water levels in parts of Kings and Queens Counties were lowered to as much as 35 feet below sea level. Sea water intrusion resulted. Ground water supplies in the most highly urbanized areas of these counties have been abandoned and water is supplied from the New York City water system. Because of the high rates of pumping and the extension of sewer systems removing a recharge source, the hydraulic imbalance persists and the threat of intrusion is continuing.

One of the most intensively studied coastal aquifers in the continental United States is the Biscayne Aquifer in southeastern Florida. It is an unconfined aquifer of limestone and calcareous sandstone extending to an average depth of 30 m below sea level. Field data indicate that the saltwater front undergoes transient changes in position under the influence of seasonal recharge patterns and the resulting water-table fluctuations (Freeze and Cherry, 1979). The entire coastal zone of the Biscayne aquifer has been affected by sea water intrusion (Klein and Hull, 1978). Most of the saltwater intrusion in the Miami area took place before 1946 when canal flow was virtually uncontrolled and ground-water levels were greatly lowered. Intrusion was halted during subsequent years as a result of installation of control structures in canals by the local and state water-control and the water-management agencies. A re-advance of saltwater occurred during the prolonged dry season of 1970-71 in the Miami and south Dade County areas.

California is another state which has experienced ground water quality degradation from sea-water intrusion. A study by the Department of Water Resources of the Resources Agency of California (1975) evaluated the status of sea-water intrusion in 263 ground water basins along California's coast. The research found:

(1)   all larger coastal ground water basins supply significant amounts of water;

(2)   intrusion is known to exist in 14 basins;

(3)   intrusion is suspected in 14 more basins;

(4)   degraded ground water of unknown origin exists in 64 basins;

(5)   no intrusion is apparent in 34 basins; and

(6)   no data is available for 137 basins.

The 1975 report on sea-water intrusion in California was based on field work conducted during 1970 and 1971, as well as a survey conducted nearly twenty years earlier (1953-1955). The 1953-1955 survey listed nine ground water basins that had been intruded by sea water, while the 1970-1971 survey identified 14 basins that had been intruded by sea water. The five additional basins are all small alluvium-filled valleys where intrusion appears to be confined to the immediate coastal portion.

Three basins were identified in the 1953-1955 survey as having a "most serious intrusion problem": Santa Clara Valley (San Francisco Bay Area); West Coast Basin (Los Angeles County); and the East Coastal Plain Pressure Area (Orange County). Encroachment of sea water in these three basins has since been partially arrested by local agencies that carried out ground water management plans developed cooperatively by the agencies and the Department of Water Resources. These efforts combined several procedures to slow the rate of intrusion: in Northern California, water taken from the South Bay Aqueduct of the State Water Project was spread and ground water pumping was reduced; and in Southern California, Colorado River water was used for injection, ground water pumping was reduced, and local and imported water was used for artificial recharge (Department of Water Resources, 1975).

Three ground water basins were identified in the 1953-1955 survey as having a "critical sea-water intrusion problem". These were the Pajaro Valley (Santa Cruz and Monterey Counties); the Salinas Valley Pressure Area (Monterey County); and the Santa Clara River Valley-Oxnard Plain Basin (Ventura County). Sea-water encroachment has continued unabated in these basins since the 1953-1955 survey (Department of Water Resources, 1975). Local agencies are well aware of the problem and over the years have conducted studies both independently and in cooperation with the State Department of Water Resources and the Water Resources Division of the U.S. Geological Survey.

Three ground water basins were identified in the 1953-1955 survey as having a "continuing sea-water intrusion problem". These were the Petaluma Valley (Marin and Sonoma Counties); the Napa-Sonoma Valley (Napa and Sonoma Counties); and the San Luis Rey Valley-Mission Basin (San Diego County). The current magnitude of sea-water intrusion in these basins is not clearly defined, but the situation can become acute if ground water extractions are increased (Department of Water Resources, 1975).

The 1970-1971 survey identified 14 areas of "suspected intrusion". All are small alluviated basins, except for two relatively large areas: the Suisun-Fairfield Valley (Solano County) and the Sacramento-San Joaquin Delta (Sacramento, San Joaquin, and Contra Costa Counties) (Department of Water Resources, 1975).

Most coastal states have experienced (in various levels of seriousness) sea-water intrusion. The following coastal states were reported in the 1983 National Water Summary (U.S. Geological Survey, 1984) as having saline ground water problems in coastal basins: California, Connecticut, Delaware, Florida, Georgia, Hawaii, Louisiana, Massachusetts, Mississippi, New Jersey, New York, North Carolina, South Carolina, Texas, and Washington. More information on these states can be found in Chapter 4.

Leaking Wells

Disposal of waste by utilizing injection wells has been practiced in many parts of the United States. The waste injected includes industrial waste, sewage effluent, spent cooling water, irrigation drainage, storm water, and brine from oil and gas activities. Of wells used for disposal of industrial wastes and municipal sewage in saline aquifers, few failures have been reported. There are, however, a number of documented cases of severe ground water contamination for shallow wells completed in potable water aquifers, frequently from the illegal use of wells for the disposal of various types of hazardous wastes.

Brine disposal from oil and gas production, on the other hand, has been a major cause of ground water contamination in areas of intense petroleum exploration and development. The principal problem has been related to the long-term practice of discharge to unlined pits, which stem from days when there was very little regulation of oil exploration and development. Presently, the major problem is discharge of saline water from abandoned oil and gas wells rather than disposal of waste brine through injection or secondary recovery wells at active petroleum recovery fields.

The following subsections delineate the various practices and their effects on the ground water environment in relation to salt water intrusion. Further, waste disposal through wells and brine disposal from petroleum exploration and development are elaborated with a few case studies to illustrate the range and nature of the contamination.

Mechanisms of Contamination

The major concern of subsurface injection through wells is the potential contamination of usable ground water by the following mechanisms (Miller, 1980):

(1)   direct emplacement into potable water zones;

(2)   escape into a potable aquifer by well-bore failure;

(3)   upward migration from receiving zone along outside of casing;

(4)   leakage through inadequate confining beds;

(5)   leakage through confining beds due to unplanned hydraulic fracturing;

(6)   leakage through deep abandoned wells;

(7)   displacement of saline water into a potable aquifer;

(8)   injection into salaquifer eventually classified as a potable water source; and

(9)   migration to potable water zone of the same aquifer.

The greatest potential for contamination occurs in the immediate vicinity of the well bore, where the natural geological structures relied upon to contain the waste have been breached by the well construction. Cement grout is forced around the casing to reinforce the geological structure, which is, however, not always as impenetrable as the original natural structure. Coupled with the acidic nature of the waste which corrodes the casing and the excessive injection pressure which can rupture the casing joints and crack the cement envelope, the waste is likely to move out into the shallower and possibly potable water-bearing zone. Excessive injection pressure can cause leaks at the well head itself, allowing effluent to flow across the land surface, perhaps contaminating ground water and surface water on the way.

Other wells in the vicinity of an injection well drilled to or through the receiving formation, such as unplugged, abandoned oil and gas wells, can act as conduits for the migration of fluid. Further, it is known that injection can reverse the hydraulic relationship. In such an instance, wells with corroded casing, or leaky wells, which have been transferring shallow fresh water downward into the deeper saline formations, will now transfer saline water upward to the potable zone.

Waste Disposal Through Wells

The greatest attention in existing literature has been given to deep well disposal of industrial and municipal effluents. As of 1980, a total of 322 such wells had been constructed in 25 states, with 209 being in operation (Miller, 1980). Due to a strict regulation and permit system generally enforced by public agencies, few failures have been reported. Further, based on reviews of the status and impact of waste injection practices in the United States and Canada, Warner and Orcutt (1973) have concluded that documented cases of even minor disposal system failure and related contamination of surface and near surface waters are rare.

Irrigation and storm water drainage wells and septic tank effluent wells total about 15,000 in Florida, Oregon, and Idaho alone. On Long Island, New York, approximately 1,000 diffusion wells inject about 80 million gpd from air conditioning or cooling systems into two of the principal aquifers tapped for public supply (Miller, 1980). Thousands more are used for disposal of storm water runoff.

Brine Disposal from Petroleum Exploration and Development

Most oil field brines today are returned to oil producing zones or deep saline aquifers through old production wells or brine injection wells. This is done either for the purpose of water flooding or secondary recovery (reinjection of brine into the oil producing zone to increase reservoir yields), or just as a disposal method. Numerous problems are, however, inherent within the method. Successful disposal requires the availability of permeable formations sufficiently thick to accommodate the amount of brine produced. These formations must lie below fresh water zones and be separated from them by impermeable layers that will prevent the upward migration of the mineralized water, a condition that is not found everywhere.

The mechanisms by which brine disposal or secondary recovery wells (water flooding) can contaminate fresh water aquifers are more complex than those for evaporation pits. Improper disposal well contruction may be intentional to save money. Such an occurrence as injection of brine into a fresh water formation by a well drilled only to a shallow depth is not uncommon (Miller, 1980). More common phenomenon include improper sealing or cementing of the casing, thus allowing brine to travel upward in the annular space between the casing and the bore hole to contaminate shallow zones; and use of easily corrodible and leakable casing.

Improperly plugged, abandoned wells and test holes are excellent conduits for the migration of brine. Where either hydrostatic pressure or pressure caused by secondary recovery operations is sufficient, brines are pushed upward in these holes to be carried to the fresh water zones; even if the well is plugged at the land surface, considerable leakage of brine into fresh water zones below the plug can occur.

Dry holes are another significant source of contamination. In 1973, more than 24,000 miles of holes were drilled on shore, of which nearly 10,000 failed to yield oil and were abandoned for the most part (some may have been used for brine disposal). Over the years, the total number of dry holes undoubtedly is on the increase; and though at present dry holes are generally plugged, those drilled in the past were commonly left open.

Upconing

Upconing is a term that describes the situation where a producing well is located close enough to saline water underlying fresh water and pumped at a rate sufficient to cause the salt water to be drawn up into the well in an upward shaped cone or mound. As indicated, the proximity of the bottom of the well in relation to underlying saline water, and the rate at which the well is pumped for the particular formation characteristics, influence the extent of upconing that will occur. Overpumping of a well can then be taken to mean that for the given hydrogeologic system a well is produced at a sufficient rate to draw up underlying saline water. The sequence of the process of upconing under unconfined aquifer conditions is illustrated in Figure 5 (U.S. Environmental Protection Agency, 1973). An estimated two thirds of the continental United States is underlain by saline ground waters that could intrude on fresh water supplies as a result of upconing. A related type of intrusion that can also occur is lateral movement of saline waters due to overpumping of a well. Lateral saline water intrusion can occur in coastal areas, as was discussed above, and can also occur inland under certain conditions. Both types of intrusion are described in this section.

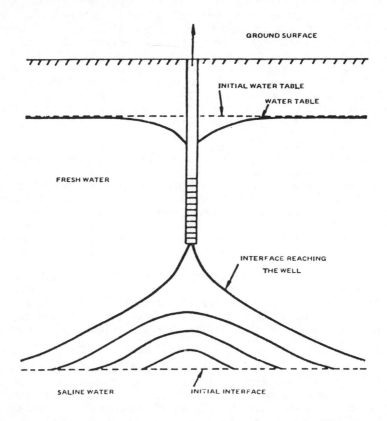

Figure 5:   Schematic Diagram of Upconing of Underlying Saline Water to a Pumping Well (U.S. Environmental Protection Agency, 1973)

In addition to overpumping, upconing of saline water can occur as a result of dewatering such as from quarries or other excavations. This could occur where saline and fresh ground waters are connected and the dewatering results in vertical migration. This phenomenon is illustrated in Figure 6 where saline water intrusion results from a lowering of the water table (U.S. Environmental Protection Agency, 1973).

Even though upconing is a complex phenomenon, criteria have been developed for the design and operation of wells to skim fresh water from above saline water.   These criteria include optimum well location, depth, spacing, pumping rate, and pumping sequence (U.S.   Environmental Protection Agency, 1973; and Wilson et al., 1976).

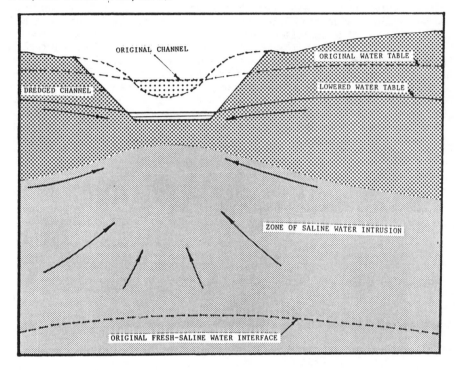

Figure 6: Diagram Showing Upward Migration of Saline Water Caused by Lowering of Water Levels in a Gaining Stream (U.S. Environmental Protection Agency, 1973)

An interesting example of upconing has been described by Mattox (1970) for the area of Chaves County, New Mexico.   A geological survey of the area was conducted by the U.S. Geological Survey in cooperation with the State Engineer.   This survey established that saline water intrusion was occurring due to overpumping.   This intrusion included both lateral and vertical movement within the artesian aquifer.   Encroachment is continuing and is a serious problem in irrigated areas and for the City of Roswell.   The patterns and magnitude of pumping, the presence of interconnected zones of preferential permeability, and the presence of fault zones in the aquifer are all factors which influence the encroachment pattern.   Modifications of pumping patterns have resulted in some trends toward improved water quality if seasonal declines of artesian head are not severe.

A second example of saline water intrusion due to overpumping has been described by Rollo (1969) for the Baton Rouge area of Louisiana.   Monitoring wells were used to determine the hydraulic gradients in the principal aquifers of the area and to collect samples to trace the migration of saline water.   Salty water intrusion in some of the aquifers was documented and the rate of migration estimated based on current pumping rates. Controls, including the use of discharge barrier wells, were also suggested.

In addition to these examples, Newport (1977) surveyed the occurrence of salt water intrusion in the United States. His survey included upconing and lateral intrusion of saline water due to overpumping. The results are shown in Table 2 for each of these two types of intrusion as reported by Newport (1977). Instances of ocean water and tidal water intrusion are not included in this table. According to this report, upconing and/or lateral intrusion occurs in 22 of the continental states. Of the sixteen states that indicated upconing is a problem, it was only suspected in three states, and in all cases the problem was attributed to overpumping. Lateral intrusion was reported to occur in ten states (four states reported both upconing and lateral intrusion occurring) from such sources as saline streams or lakes including the Great Salt Lake area in Utah (Newport, 1977).

Table 2:   States Reporting Upconing or Lateral Intrusion of Saline Water
(modified from Newport, 1977)

State Reporting Upconing	States Reporting Lateral Intrusion
Alabama - Marango County coastal plain	Alabama - Mobile River
Arkansas - Southern	Arkansas - Eastern
Florida - Cocoa Beach Cape Canaveral and Hendry County - Southwest	
Illinois - Suspected	Illinois - Suspected
Indiana - Suspected in Mt. Vernon - West Franklin	
Iowa - Suspected	
	Kansas - Stream flow infiltration potential
	Kentucky - Stream flow infiltration
	Louisiana - Baton Rouge
Michigan - Various	
Minnesota - Northwest	
Missouri - Potential	

Table 2: (Continued)

State Reporting Upconing	States Reporting Lateral Intrusion
Nebraska - Potential in Northeast and East	
	Nevada - Potential
New Mexico - Various	
North Dakota - Red River Valley	
South Carolina - Paris Island	
South Dakota - Potential in Black Hills and various other areas	
Texas - Galveston Texas City	
Utah - Western	Utah - Great Salt Lake area and Western
	Washington - Saline Lakes in Grant County
	Wisconsin - Various

Natural Salinity

The term "saline groundwater" is defined in many different ways by the professionals and organizations that work in this area. These definitions are easier to correlate with one another when it is noted that although there is more than 40 chemical types of mineralized waters, the sodium chloride type is the most common in every region of the United States, comprising up to 35 percent of all analyses (Feth, et al., 1965). In almost all waters containing more than 20,000 ppm, the dominant anion is chloride (Feth, 1981). The major exceptions to this rule are waters associated with saline seeps and the oxidation of sulfide minerals in mines in which sulfate is the dominant anion. Chloride (Cl^-) is generally conservative, traveling through the hydrologic cycle with less involvement in oxidation-reduction reactions, adsorption, or the life processes of plants or animals than is true of most common ions. Therefore, Cl^- has been considered to be a good tracer (Feth, 1981), and a good standard against which changes in the abundance of other ions can be evaluated (Blatt, Middleton, and Murray, 1980).

Table 3 shows the concentrations of chloride and other salinity related ions in subsurface and some surface waters (Craig, 1970).   One possible explanation for the ions in the subsurface waters is that they are juvenile waters and magmatically derived from fluids.   Fluids that were originally believed to be genuinely juvenile, however, such as those geothermal brines obtained in wells near the Salton Sea in Southern California (Table 3, No. 9), are now felt to be primarily, if not entirely, local meteoric water (Craig, 1970). A large number of volcanic gases have been analyzed and, in nearly every case, the degree of meteoric contamination is felt to be great (Craig, 1970).   This does not seem to be a major source of salinity in subsurface water.

Numerous investigators have noted that waters within sedimentary strata, especially in oil fields where work has been concentrated, become increasingly saline with increases in depth.   In general, the sequence noted is sulfate-rich waters near the surface, more saline bicarbonate-waters at an intermediate level, and yet more concentrated chloride waters at greater depths (Craig, 1970). Since the formations of many thick sedimentary sequences are marine in origin, especially those in oil fields, virtually all authors have considered the waters to have originated in part as sea water (Craig, 1970).

It has been pointed out, however, by the comparison in Table 4 between seawater and formation waters made by Blatt, Middleton, and Murray (1980), that subsurface waters are not merely entrapped sea water or meteoric infiltrations, nor their concentrates.   This is also noticeable from Table 3. These waters appear to have undergone selective chemical reactions which change the proportions of dissolved solids. This assumption is supported by the present knowledge of chemical diagenesis of waters in sediments.

There are several mechanisms by which entrapped water can be altered to the saline water and brines found in the subsurface. One of these is the solution of sediments and rocks.   This seems to be a very unimportant mechanism if consideration is given to the chloride content of most minerals. Estimates suggest that in igneous rocks concentrations are less than 500 ppm and, in sedimentary rocks, excluding evaporites, from 10 or less to about 1,500 ppm. In a few individual minerals such as sodalite, apatite, mica, and hornblende, where chloride replaces hydroxyl, and in scapolite, the concentration may reach 50,000 ppm or more (Feth, 1981).

Evaporite minerals, on the other hand, could be the source of a large part of salinity in water. The most common evaporite, halite (NaCl), is about 60% Cl^-. About one fourth of the continental areas of the world may be underlain by evaporites, and about 60 percent of these areas are underlain by salts that contain Cl^-. Most saline deposits are in the northern hemisphere where more than one-half the areas underlain by nonevaporite sedimentary rocks also are underlain by evaporites (Feth, 1981).

Specific examples of the solution of subsurface evaporites have been cited by White et al., (1963) for several waters in Table 3 (Nos. 4, 5, and 6). No. 4 is a sodium carbonate brine in contact with Eocene trona deposits; the low content of all elements except Na, carbonates, and Cl are deemed to support an origin by salt solution in meteoric water. No. 5 is a brine formed by solution of the Timbalier Bay salt dome in Louisiana; note the low K, Br, and B contents. No. 6 is from New Mexico, probably from the "lower salt series" of the gypsum and anhydrite-bearing Castile formation (Craig, 1970).

Table 3:   Ion Concentrations in Natural Waters (concentrations in parts per million) (Craig, 1970)

	Ocean water	Oilfield brine California	Oilfield brine Wyoming	Seep from a Trona deposit Wyoming	Oilfield brine Timbalier Bay Saltdome, La.	Test well brine New Mexico	Mine water Comstock Lode	Mine water Coal Mine, W. Va.	Geothermal well Salton Sea, Calif.	Salton Sea	Salt Lake
	1	2	3	4	5	6	7	8	9	10	11
$CO_3^= + HCO_3^-$	143	112	201	102300	147	1380	-	-	150	232	180
$SO_4^=$	2712	1590	6650	167	2.6	222000	78600	5940	5.4	4139	13590
F^-	1.3	0.3	6.0	18	1.1	-	-	-	15	1.6	-
Cl^-	19000	1300	596	21300	89700	16800	120	174	155000	9033	112900
Br^-	65	7.7	0.7	22	86	-	-	-	120	-	-
Ca^{++}	400	501	216	3.7	2600	97	1160	314	28000	505	330
Mg^{++}	1350	103	99	4.6	1060	38300	6240	224	54	581	5620
Na^+	10500	1020	3250	84700	51400	43900	498	896	50400	6249	67500
K^+	380	9.8	78	158	193	2090	469	18	17500	112	3380
Fe total	0.01	-	0.8	-	-	-	576	1210	2090	-	-
SiO_2	6.4	59	36	47	25	-	-	118	400	20.8	-
Total Dissolved Solids	34475	5050	13300	209000	145000	325000	97700	8894	258360	20900	203490

Table 4:   Comparison Between Dissolved Species in Modern Seawater and in Formation Waters (Blatt, Middleton, and Murray, 1980)

Characteristic	Modern seawater	Range in formation waters	Median in formation waters	Direction of change
pH	7.9-8.2	3-11	9	Increase
Eh	-0.2 to +0.4v	-0.4 to +0.7v	?	?
Chlorinity	0.535m	up to 7.61m	2.48m	Increase
mNa/mCl	0.86	0.15-0.95	0.72	Decrease
mMg/mCl	0.097	0.003-0.117	0.032	Decrease
mSO$_4$/mCl	0.052	$7x10^{-6}$-0.013	0.002	Decrease
mK/mCl	0.019	$1x10^{-3}$-0.064	0.008	Decrease
mCa/mCl	0.019	0.008-0.34	0.113	Increase
mHCO$_3$/mCl	0.004	$2x10^{-5}$-0.024	$3.2x10^{-3}$	Decrease
mCO$_3$/mCl	$0.7x10^{-4}$	?	$4.6x10^{-4}$	Increase
mSr/mCl	$1.7x10^{-4}$	$1x10^{-3}$-0.013	$1.9x10^{-3}$	Increase
mH$_4$SiO$_4$/mCl	$1.9x10^{-5}$	$3x10^{-5}$-$5x10^{-4}$	10^{-4}	Increase
mCa/mMg	0.19	0.56-26.92	3.53	Increase
Percentage "total CO$_2$" as CO$_3$ ion	0.7%	?	10%	Increase

Evaporites occur not only in discrete, mappable beds, but also as grains and cements in clastic sedimentary rocks (Feth, 1981). A related mechanism is leaching and incongruent dissolution of detrital minerals in clastic rocks (Blatt, Middleton, and Murray, 1980). Of particular importance in respect to salinity is the removal of calcium, sodium, and potassium from feldspars. Critical diffuse salt source areas occur in much of the semi-arid and arid parts of the western states. Over one half of the average annual 10.7 million tons of salt in the Colorado River is from natural diffuse salt areas (U.S. Department of Agriculture, 1975).

Another mechanism by which subsurface water can be altered is by the change in partial pressures of gases in the water in response to organic and inorganic processes. Sulfate, for example, is converted to sulfide by bacteria during the formation of sulfur deposits with the release of oxygen into the water; also carbon dioxide is released from petroleum accumulations (Blatt, Middleton, and Murray, 1980). These obviously effect the concentrations of sulfate, carbonate, and calcium in the waters, but their significance is questionable because they are so localized.

The explanation for the chemical changes that is currently cited most often is the "salt sieving" process. This theory was first introduced by deSitter (1947). Basically, as fine silts and clays are deposited, the water and aqueous ions that are associated with them are free to move unimpeded. As compaction occurs and the clay minerals are placed very close together, the fixed negative charges on the clays come close enough to repel the aqueous anions (such as Cl-) while permitting the passage of uncharged water molecules. Cations may pass through this clay "membrane" in a stepwise fashion, moving from one cation exchange site to another. As the cations move up through the membrane, hydrogen ions move down to maintain electric neutrality on the high pressure side. This also effectively lowers the pH on the high pressure side.

In 1963, Bredehoeft et al., amended this theory by including the hydraulic head in confined aquifers as the driving force moving the water and ions upwards through the membrane. With this addition this theory can be used to explain brine waters that are being "created" at this time from meteoric waters, as well as those created at the time of deposition. This theory also explains many of the following phenomena observed in subsurface waters: increasing acidity and salinity with depth; changes in ion ratios that coincide with their relative mobilities through clays; and the occurrence of brines in areas away from the sea and with no salt deposits.

Saline Seeps

Saline seeps are defined as recently developed saline soils in nonirrigated areas that are wet some or all of the time, often having white salt crusts (Miller et al., 1980). It is a place where once productive land becomes nonproductive due to the seepage of saline ground water. Seepage of saline ground water was recognized in Montana in the 1800's. These salt spots were present before the land was first cultivated. They do not spread and are not included in the modern definitions of saline seep. All definitions stipulate that the wet salty area must appear where historically it did not exist before (Thompson and Custer, 1976).

Early in the development of a seep the signs are subtle. The soil surface may remain damp longer than normal after a rain (2-3 days). During plowing, the tractor engine may labor unusually as the implement passes over moist soil. The soil may show localized clodiness. The clods may have a very thin bitter-tasting salt crust on them when dry. A moisture probe can be pushed into the ground without resistance and free water can often be seen on the tip of the probe when it is withdrawn (Thompson and Custer, 1976). Crop production declines; however, a localized high yield may occur the year before, due to excess water in the area before the salt builds up (Thompson and Custer, 1976). Initially there are problems with salt-tolerant weeds. There is often a weed succession in the seep; for example, Kochia, Russian Thistle, Foxtail Barley, and Glasswort. Often the plants are zoned, but this has not been studied in detail (Thompson and Custer, 1976).

In 1974, available analyses of water collected near Fort Benton and Sidney, Montana, Mott, North Dakota, and Lethbridge, Alberta, strongly suggested that in addition to losing thousands of acres of valuable farmland to saline seeps, mineralized water was rapidly contaminating nearby reservoirs, streams, and shallow aquifers (Miller et al., 1980). Reservoirs south of Fort Benton which used to support trout were barren. Farmers reported that calves drinking from stock ponds and streams in affected areas developed scours (a form of dysentery), staggers, and, occasionally, blindness. They also reported deterioration of domestic supplies (Thompson and Custer, 1976).

The effect of saline seep on wildlife habitat and population is uncertain. The salt tolerant weeds at the fringes of the seeps provide food and cover where none existed before (Bahls and Miller, 1975). There is a major concern, however, especially among wildlife biologists in Montana, that saline seeps may eventually destroy or greatly reduce the habitat of upland game birds in Montana's grain belt and the nesting areas of migrant waterfowl in Montana's prairie potholes (Bahls and Miller, 1975).

Saline seeps are usually found in areas of thin alluvial or glacial aquifers. The basic ground water systems in these areas are shallow, locally-derived, and perched. This has been evidenced by several factors:

(1)   when drilling, the bit usually penetrates the saturated zone and passes into a powder dry shale in less than 60 ft (Thompson and Custer, 1976);

(2)   the water table reflects the topography (Custer, 1979);

(3)   there is no relation between saline seep distribution and known geological faults where these relations have been studied (Custer, 1979); and

(4)   the static water level in saline seeps responds rapidly to precipitation (Custer, 1979; and Thompson and Custer, 1976).

The recharge and discharge areas are often contained in less than one square mile (Thompson and Custer, 1976). The seep develops in response to the elevation of the water table. The process starts by movement of water beneath the root zone but above the shallow impermeable layers, thereby forming the local ground water flow system. The flow system moves saline water downslope to the discharge area (seep), where it evaporates, depositing the salt on the surface (Miller et al., 1980). Figure 7 shows the most commonly pictured system of the internally drained upland recharge area and the internally drained lowland discharge area (Thompson and Custer, 1976). This type of system may develop on glacial till, colluvium, fluvial gravels, or bed rock.

Seeps break out on slopes as well as in bottom lands. This can be the result of a lateral downslope decrease in permeability. As shown in Figure 7, in some areas there is a permeable aquifer, such as a clinker or coal bed, which is truncated by erosion forming an outlet on the hill below the recharge area. Such a system may also be covered by glacial till or by a thin mantle of soil (Thompson and Custer, 1976). A similar result can be created by a change in the gradient of the perching unit, which can act to temporarily pond the ground water and raise the water table locally (Thompson and Custer, 1976).

Although saline seep systems are thought usually to have a very local recharge area, intermediate flow systems may develop in some areas according to Thompson and Custer (1976). For example, deep glacial and preglacial river channels as much as 100 m (300 ft) deep may be covered with glacial till and outwash material. The channels may have relatively large recharge areas and may develop intermediate flow systems as shown in Figure 7. Discharge may be enhanced by lateral discontinuities in permeability or gradient (Thompson and Custer, 1976).

As mentioned before, saline seeps increase during periods of high precipitation; however, they also occur in areas with as little as 12 inches per year. In Conrad, Pondera County, Montana, the average precipitation has decreased from 13.8 inches in the period 1940-51 to 11.6 inches in the period 1961-70 (Thompson and Custer, 1976). Saline seep is still a problem in the

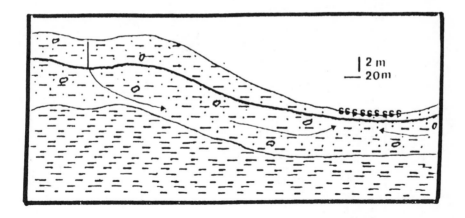

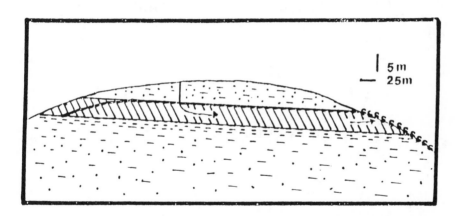

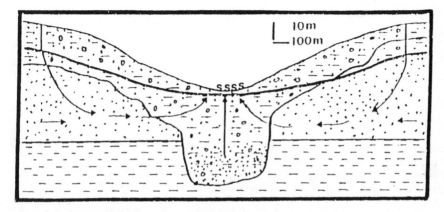

Figure 7: Cross Section of Groundwater Flow to a Saline Seep
(Thompson and Custer, 1976)

Explanation

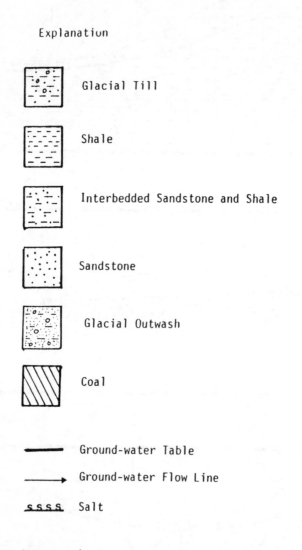

Glacial Till

Shale

Interbedded Sandstone and Shale

Sandstone

Glacial Outwash

Coal

Ground-water Table

Ground-water Flow Line

ssss Salt

Figure 7: (Continued)

Conrad area, which indicates that although rainfall contributes to it, it is not the primary cause. The primary cause appears to be farming practices in general, and the alternate crop-fallow system in particular (Thompson and Custer, 1976; Miller et al., 1980; and Donovan, Sonderegger, and Miller, 1981).

In nature the grasses, forbes, and shrubs of the prairie make efficient use of water and restrict excessive vertical percolation. This efficient water use implies a permanently expanded root system at all depths in the soil profile.

Annual grasses and forbs, like small grain crops, obtain their water only from the upper soil levels; their root systems are shallow, expanding and functioning only during the growing season. Perennial grasses and shrubs, absent from most cropping systems, use water that escapes from shallow-rooted plants into the subsoil; their root systems are deeper, permanently expanded, and functional over a longer period (Bahls and Miller, 1975).

During the summer fallow period in the crop fallow rotation, the soil is kept bare of all vegetation to enhance the soil moisture status for the next crop. Only about 27 percent of the precipitation is stored in the soil during the fallow period; the remainder is lost to evaporation and to deep percolation (Thompson and Custer, 1976). Research indicates that fallow areas can undergo a water-table rise of 1 to 15 feet during years of average or above-average spring precipitation. The water levels gradually decline during the rest of the year but normally do not reach the low of the previous year. As a result, excess water accumulates through the years, causing expansion of the saline seeps during each succeeding wet cycle (Miller et al., 1980).

Although the concentration of water soluble salts in the soil profile and substratum are quite variable, the chemical composition of saline-seep water is remarkably uniform. During the evolution of a typical saline seep, the ground water quality changes from calcium bicarbonate type of water with relatively low total dissolved solids (TDS) (1,500 to 3,000 mg/l) to a sodium-magnesium sulfate type of water with high TDS (4,000 to 60,000 mg/l) (Miller et al., 1980). In addition to the high TDS, saline-seep water commonly contains much higher concentrations of nitrates and trace metals, such as aluminum, iron, manganese, strontium, lead, copper, zinc, nickel, chromium, cadium, lithium, and silver. Most of these elements are rarely detected in water samples from other areas (Bahls and Miller, 1975). Because of these factors and the low chloride concentration, saline seep water is readily distinguished from deep subsurface brines.

The saline seep problem is endemic but not limited to the entire Northern Great Plains Region (Montana, North Dakota, South Dakota, Alberta, Manitoba, and Saskatchewan). Hardly recognized 30 years ago, it has now taken roughly 2 million acres out of this region (Miller et al., 1980). Figure 8 shows the area of potential saline seep development (Bahls and Miller, 1975). In 1980, surveys indicated that saline seep had taken roughly 200,000 acres of Montana's dryland from agricultural production, but Miller, et al., (1980) believe that that figure may be somewhat low.

On the average, the acreage affected by saline seep is increasing at an annual cumulative rate of 10 percent per year relative to the first year of measurement (Thompson and Custer, 1976). Expansion of seep areas by 20 to 200 percent in wet years is not uncommon, whereas expansion of only a few percent may occur in dry years. Currently, seep development is especially rapid in areas where glacial till is less than 50 feet thick. Excess water is probably building up also in extensive areas underlain by greater thicknesses of till, but as yet the buildup is not evident at the surface (Miller et al., 1980). All areas with local shallow water tables where the water percolates through saline or alkali soils or strata have the potential for problems with saline seep. Areas of dryland farming where the crop-fallow system is in use should be especially conscious of this seep potential.

Currently, soil moisture management through crop selection and farming

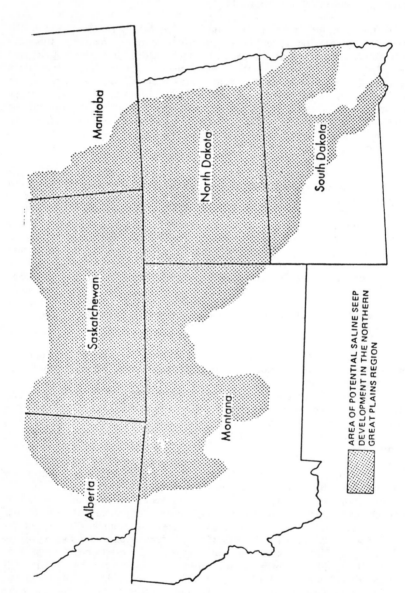

Figure 8:   Northern Great Plains Region, Showing Area of Potential Saline
Seep Development (Bahls and Miller, 1975)

methods is probably the only practical means of controlling saline seep. The nonfarming practices do not appear amenable to widespread application (Bahls and Miller, 1975). The most obvious method for control is to crop every year instead of every other year. Near Fort Benton, Montana, 4 years of alfalfa in the recharge area of a previously summer-fallow saline seep produced a water table drop of 8 ft. Alfalfa planted on the recharge area of a previously summer-fallowed saline seep near Sidney, Richland County, Montana, has led to a decrease in the size of the seep. Continuous cropping has contributed to the stabilization of a seep in an area where others are growing near Sidney and near Molt, Stillwater County, Montana (Thompson and Custer, 1976).

Nonfarming methods can be used to control the flow of saline water out of a seep area as well as the flow of fresh water into the recharge area. These methods include drainage of water-filled surface recharge potholes, direct tile drainage of saline seeps, land grading, and phreatophytes.

SELECTED REFERENCES

Bahls, L.L., and Miller, M.R., "Ground Water Seepage and Its Effects on Saline Soils-Completion Report", MUJWRRC Report No. 66, October, 1975, Montana University Joint Water Resources Research Center, Montana State University, Bozeman, Montana.

Ballentine, R.K., Reznek, S.R., and Hall, C.W., "Subsurface Pollution Problems in the United States", Technical Studies Report TS-00-72-02, May, 1972, U.S. Environmental Protection Agency, Office of Water Programs, Washington, D.C.

Blatt, H., Middleton, G., and Murray, R., Origin of Sedimentary Rocks, 2nd Ed., Prentice Hall, Englewood Cliffs, New Jersey, 1980.

Bredehoeft, J.D. et al., "Possible Mechanisms for Concentration of Brines in Subsurface Formations", American Association of Petroleum Geologists' Bulletin, Vol. 47, 1963, pp. 257-269.

Craig, J.R., "Saline Waters: Genesis and Relationship to Sediments and Host Rocks", in Ed., Mattox, R.B., Saline Water, Proceedings of the Symposium on Groundwater Salinity, 46th Annual Meeting of the Southwestern and Rocky Mountain Division of the American Association for the Advancement of Science, Contribution No. 13 of the Committee on Desert and Arid Zones Research, Las Vegas, Nevada, 1970.

Custer, S.G., "Regional Identification of Saline-Seep Recharge Areas, Stillwater and Yellowstone Counties, Montana", MWRRC Report No. 102, November, 1979, Montana Water Resources Research Center, Montana State University, Bozeman, Montana.

Department of Water Resources, The Resources Agency, State of California, "Sea-Water Intrusion in California: Inventory of Coastal Ground Water Basins", Bulletin No. 63-5, October, 1975.

de Sitter, L.U., "Diagenesis of Oil Field Brines", American Association of Petroleum Geologists' Bulletin 31, 1947, pp. 2030-2040.

Donovan, J.J., Sonderegger, J.L., and Miller, M.R., "Investigations of Soluble

Salt Loads, Controlling Mineralogy, and Factors Affecting the Rates and Amounts of Leached Salts", MWRRC Report No. 120, September, 1981, Montana Water Resources Research Center, Montana State University, Bozeman, Montana.

Erinakes, D., personal communication, U.S. Soil Conservation Service, Ft. Worth, Texas, January, 1985.

Feth, J.H., "Chloride in Natural Continental Water--A Review", Geological Survey Water-Supply Paper 2176, U.S. Geological Survey, Washington, D.C., 1981.

Feth, J.H. et al., "Preliminary Map of the Conterminous U.S. Showing Depth to and Quality of Shallowest Ground Water Containing More Than 1000 Parts Per Million Dissolved Solids", Atlas HA-199, 1965, U.S. Geological Survey, Washington, D.C.

Freeze, R.A, and Cherry, J.A., Groundwater, Prentice-Hall, Inc., Eaglewood Cliffs, New Jersey, 1979.

Ghyben, B.W., "Nota in Verbanment de Voorgenomen Put Boring Nabij Amesterdam" (Notes on Probable Results of the Proposed Well Drilling Near Amsterdam), Tijdschr K. Inst. Ingen, Vol. 21, 1888.

Holburt, M.B., "Salinity Control Needs in The Colorado River Basin", Managing Irrigated Agriculture to Improve Water Quality, Proceedings of National Conference on Managing Irrigated Agriculture to Improve Water Quality, U.S. Environmental Protection Agency and Colorado State University, 1972, pp. 19-25.

Hubert, J.S., and Canter, L.W., "Health Effects from Ground Water Usage", NCGWR 80-17, 1980, National Center for Ground Water Research, University of Oklahoma, Norman, Oklahoma.

Jenkins, S.R., and Moore, R.K., "Effect of Saltwater Intrusion on Soil Erodibility of Alabama Marshlands", Water Resources Research Institute, Auburn University, Auburn, Alabama, July, 1984.

Klein, H., and Hull, J.E., "Biscayne Aquifer, Southeast Florida", Water-Resources Investigation 78-107, September, 1978, U.S. Geological Survey, Washington, D.C.

Lauchi, A., and Epstein, E., "Mechanisms of Salt Tolerance in Plants", California Agriculture--Special Issue: Salinity in California, Vol. 38, No. 10, October, 1984, pp. 18-19.

Maas, E.V., "Crop Tolerance", California Agriculture--Special Issue: Salinity in California, Vol. 38, No. 10, October, 1984, pp. 20-22.

Mattox, R.B., Proceedings of the Symposium on Groundwater Salinity, 46th Annual Meeting of the Southwestern and Rocky Mountain Division of the American Association for the Advancement of Science, Contribution No. 13 of the Committee on Desert and Arid Zones Research, Las Vegas, Nevada, 1970.

Miller, D.W., Ed., "Waste Disposal Effects on Ground Water: A Comprehensive

Survey of the Occurrence and Control of Ground Water Contamination Resulting from Waste Disposal Practices", Premier Press, Berkeley, California, 1980.

Miller, M.R. et al., "Regional Assessment of the Saline-Seep Problem and a Water-Quality Inventory of the Montana Plains", MWRRC Report No. 107, April, 1980, Montana Water Resources Research Center, Montana State University, Bozeman, Montana.

Newport, B.D., "Salt Water Intrusion in the United States", EPA-600/8-77-011, July, 1977, Robert S. Kerr Environmental Research Laboratory, U.S. Environmental Protection Agency, Ada, Oklahoma.

O'Leary, J.W., "The Influence of Ground Water Salinity on Plant Growth", Proceedings of the Symposium on Ground Water Salinity, Ed., Mattox, R.B., 1970, 46th Annual Meeting of the Southwestern and Rocky Mountain Division of the American Association for the Advancement of Science, Las Vegas, Nevada.

Pye, V.I., Patrick, R., and Quarles, J., Groundwater Contamination in the United States, University of Pennsylvania Press, Philadelphia, Pennsylvania, 1983.

Rollo, J.R., "Salt Water Encroachment in Aquifers of the Baton Rouge Area-Louisiana", Water Resources Bulletin No. 13, August, 1969, Department of Conservation, Louisiana Geological Survey, Baton Rouge, Louisiana.

Summers, W.K., and Schwab, G.E., "A Survey of Saline Ground Water as a Mineral Resource", in Mattox, R.B., Ed., Saline Water, Proceedings of the Symposium on Groundwater Salinity, 46th Annual Meeting of the Southwestern and Rocky Mountain Division of the American Association for the Advancement of Science, Contribution No. 13 of the Committee on Desert and Arid Zones Research, Las Vegas, Nevada, 1970, pp. 31-45.

Task Committee on Saltwater Intrusion, "Saltwater Intrusion in the United States", Journal of the Hydraulics Division, Proceeding of the American Society of Civil Engineers, Vol. 95, September, 1969.

Thompson, G.R., and Custer, S.G., "Shallow Ground-Water Salinization in Dryland-Farm Areas of Montana", MUJWRRC Report No. 79, September, 1976, Montana Universities Joint Water Resources Center, Bozeman, Montana.

Todd, D.K., "Salt-Water Intrusion and Its Control", Journal of American Water Works Association, Vol. 66, No. 3, March, 1974, pp. 181-187.

U.S. Department of Agriculture, "Erosion, Sediment, and Related Salt Problems and Treatment Opportunities", December, 1975, Soil Conservation Service, Special Projects Division, Golden, Colorado.

U.S. Environmental Protection Agency, Office of Air and Water Programs, "Identification and Control of Pollution from Salt Water Intrusion", EPA-430/9-73-013, 1973.

U.S. Geological Survey, National Water Summary 1983, 1984, Reston, Virginia.

Warner, D.L., and Orcutt, D.H., "Industrial Wastewater Injection Wells in

United States: Status of Use and Regulation", Underground Waste Management and Artificial Recharge, Ed., Braunstein, J., American Association of Petroleum Geologists, U.S. Geological Survey, International of Association of Hydrological Science, 2, 1973.

White, D.E. et al., "Chemical Composition of Subsurfgace Waters", Chapter F, Data of Geochemistry, 6th Ed., U.S. Geological Survey Professional Paper 440-F, 1963.

Wilson, J.L. et al., "Groundwater Pollution: Technology, Economics and Management", Report No. TR 208, January, 1976, Massachusetts Institute of Technology, Cambridge, Massachusetts.

CHAPTER 3

SALINITY EVALUATION, CONTROL, AND MANAGEMENT

As described in the previous chapter, salt water intrusion into fresh ground water can result from both natural conditions and many human actions. Locating and defining the magnitude of the problem is the first and most important step to controlling intrusion. Each control program must be uniquely designed depending on the type of intrusion, the specific source(s), the hydrology of the area, and the areal extent of the problem. A generalized planning approach to salt water intrusion control consists of the following five steps:

(1)   problem definition;
(2)   inventory and analysis;
(3)   formulation of alternative control plans;
(4)   comparative evaluation of control plans; and
(5)   selection and implementation of controls.

This chapter is divided into five sections which can be related to the five major planning steps on salt water intrusion control. Ground water monitoring is used to identify and define the extent of the problem and to observe trends. This is described in the first section of this chapter. The second step can involve taking an inventory of water users and use patterns when overpumping is involved and can also include the use of mathematical models to help understand and predict the movement of saline water under various conditions. The formulation of alternative controls can involve various approaches for sea water intrusion control as outlined in the third section of this chapter, or approaches for noncoastal intrusion as described in the fourth section. Water desalination is a treatment alternative, described in the fifth section, that does not control intrusion but can be used to provide water supply of the desired quality. The last section of the chapter contains a description of the legal and institutional aspects of salt water intrusion control. These aspects are basic to the selection and implementation of controls.

MONITORING

The first step in monitoring ground water for the presence of salt- water intrusion is to determine the specific purpose of the monitoring program. Four potential purposes include ambient trend monitoring to define temporal and spatial conditions in a basin or area, specific source monitoring, case preparation to gather evidence for enforcement, and research. The first two purposes, ambient trend and specific source monitoring, are most relevant to this national review of salt water intrusion and are described in more detail. Examples will be included in this section along with a description of methods used for saturated zone monitoring and vadose zone monitoring.

Before installing monitoring wells, background data should be obtained to guide the effort. This includes descriptions of the physical location and site boundaries of the areas of concern. Types of information that are useful include areal photographs, topographic maps, identification of existing monitoring and ground water data, soil borings and well logs, surface geology, and soil descriptions. This data is usually available from various agencies, including State Geological Surveys, the U.S. Geological Survey, the U.S. Soil Conservation Service, State and County Departments of Health, State and Federal Environmental Protection Agencies, State Mining and Oil Commissions, and universities.

If existing wells in the area of concern can be utilized for sampling, these need to be identified. It is important to determine the exact description of the well, including screened interval(s) and geological formations penetrated. A well may be screened across several water-bearing zones, for example, and will represent the composite quality of the ground water.

In the absence of adequate existing wells, earth electrical resistivity measurements may be used to indicate the quality of shallow ground water in an area. Stewart and Gay (1982) evaluated the use of a remote sensing technique called transient electromagnetic sounding to locate the saltwater-fresh water interface along the Gulf Coast of Florida. Penetration depths up to 500 meters were found possible, although the method is ineffective at depths less than 50 meters (other electromagnetic methods are effective in this range). Relatively sophisticated data interpretation and training are needed; even so, the method could be less costly than installing monitoring wells, and it was found to be acceptably accurate for the application tested.

Another example of the use of an earth resistivity method combined with sodium/chloride ratios is a study by the Oklahoma Water Resources Board (1975) to identify the source of salt water in terrace deposits. The surface resistivity results were compared to water quality survey results at 34 sites. It was found that the surface resistivity measurements were indicative of the presence of saline water under relatively simple geological conditions.

Several drilling methods are available for monitoring well installation. These methods can be selected for the particular types of material to be penetrated and the expected total depth of the well. Mud rotary drilling, for example, is available throughout the United States. All types of geological material can be drilled to great depths and is relatively inexpensive. Disadvantages include the use of drilling fluids which can contaminate the bore hole, and the need for electric logging to accurately locate water-producing zones. Air rotary drilling avoids the use of a drilling fluid, but difficulties can occur with cavings in water-bearing formations. These disadvantages can be partially overcome by the use of air drilling with a casing hammer, but the technology is not as readily available.

Cable tool drilling can be used in all types of formations, and samples of formations including permeabilities and approximate water quality can be obtained during the drilling. The drawbacks with this method are that it is relatively slow, costs can be large, and casing enplacement in unconsolidated formations can result in cross contamination of water-bearing zones. Auger techniques are limited to approximately 150 feet in depth, often result in collapse of water bearing formations into the borehole when the auger is withdrawn (except for hollow-stem augers), and cannot be used in hard rock. The advantages are no drilling fluids are used which can cause contamination, the drilling rigs are available, and they are relatively inexpensive.

A final method of drilling that has some applications is called jetting. This technique is fast, very inexpensive, and can be used to place shallow well points. However, this method results in the introduction of foreign water or drilling mud into the formation; it is limited to soft and shallow formations, and the diameter of the casing is usually limited to two inches.

After a borehole is created, geophysical logging of the vertical profile can be used to determine the depth to various types of strata, and it can help to identify water-bearing zones. The logging may include the use of such techniques as neutron moisture logging, gamma ray attenuation, heat dissipation sensors, resistor/capacity sensors, seepage metering, and electrical conductivity. Examples of the use of well logging to help define salt water intrusion in the coastal plain of Delaware have been described (Woodruff, 1969; and Groot, 1983). In studying the intrusion of saline ground water in Delaware aquifers during the late 1960's, Woodruff (1969) recommended that four specific investigations be conducted to further document and define the existance of encroachment. Groot (1983) reported the use of self-potential logs to interpret chloride concentrations in onshore and offshore wells in the area. He was able to determine the areas of high saline water and reach conclusions about the rate of brine movement and optimum locations for drinking water supply wells.

Basin-wide ambient trend monitoring can be used to determine if contamination is occurring over a wide area. A three-phased approach to basin-wide monitoring is shown in Figure 9. For the first phase existing information on well and spring monitoring, along with other hydrogeologic information, is analyzed to determine if additional monitoring points are needed. If new wells are installed, then the siting, drilling, casing development, along with sampling and analysis procedures, must be selected. This constitutes the second phase of the approach. Once the network is in place, then the third phase of periodic sampling, sample analysis, data interpretation, and reporting can progress. Key decisions include the number of wells to be included in the network, the spacing and location of those wells, and the optimum sampling frequency.

The other type of ground water monitoring related to salt water intrusion is associated with monitoring specific sources. This includes monitoring within aquifers at some distance from production wells, such as along the ocean coast, as well as monitoring along the saltwater/fresh water interface near production wells, for early upconing detection. The U.S. Environmental Protection Agency (1973) has described monitoring procedures for sea water intrusion control and tidal gates and locks. The monitoring requirements include observing conditions within and outside the intrusion zone, including ground water levels and chloride concentrations. At a few key locations, the vertical profile of the chloride transition zone should also be determined. The number of wells required varies with each site. For example, in a recharge line in the West Coast Basin of Los Angeles, about 30 observation wells were drilled for each mile. Parameters were normally measured every one to two months, including testing by electrical conductivity logging.

Another example of specific source monitoring was described by Larson (1981) for salt water distribution in Virginia coastal plain aquifers. He analyzed about 1,000 samples collected over 70 years from more than 700 wells. Increases in chloride levels were only local in extent and occurred in pumping wells near salty surface water sources, such as Chesapeake Bay. From this sampling effort, horizontal and vertical profiles of chloride concentrations were developed for a series of geohydrologic cross sections.

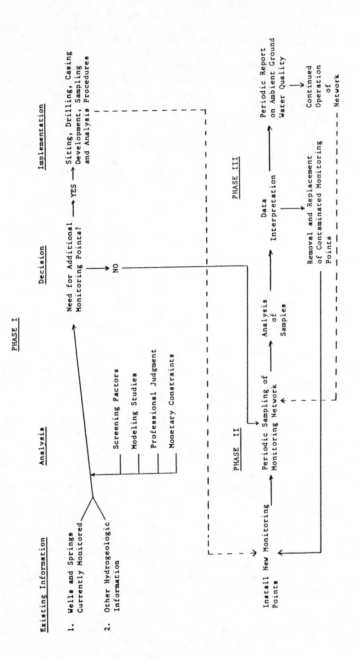

Figure 9:   Flowchart for Ambient Trend Monitoring Planning

Monitoring in the vadose or unsaturated zone could be used to detect soil pore water salinity. In addition to geophysical methods which can indicate areas of high salinity, a vacuum lysimeter can be installed to obtain a sample of the soil water by a vacuum pump. The lysimeter consists of a pipe, usually constructed with polyvinyl chloride (PVC), with a porous ceramic cup attached to one end and a polyethylene discharge line out the other end through a PVC cap for sample recovery. The lysimeter must be carefully emplaced and sampled. Saline water migration in the vadose zone could be monitored by this technique.

A specific ground water monitoring program must be developed for each specific source or ground water basin to be investigated. The objective of the monitoring effort is to collect and analyze ground water quality, geologic, and hydrologic data to enable the development of an understanding of the nature and extent of any intrusion problems that are occurring or are likely to occur. This may involve data collection for one or more of the following purposes:

(1)   provision of background information on quality;

(2)   detection of quality trends;

(3)   identification and assessment of the sources and causes of pollution;

(4)   planning, including the use of models;

(5)   establishment of water quality standards and effluent limitations;

(6)   formulation of other regulatory control and management actions necessary to protect quality;

(7)   compliance;

(8)   enforcement; and

(9)   reporting.

The relative emphasis placed on each of these purposes will vary depending on the extent of ground water development that already exists. Several types of data may be needed, including water quality, geologic characteristics, hydrology, hydraulic characteristics, land use, water resource development and use, demographic and economic conditions, and waste generation and disposal. Since ground water is not static because of natural processes and man's activities, the monitoring effort should be continually evaluated for effectiveness.

MODELING SALT WATER INTRUSION

This section contains a brief description of some widely used modeling techniques in ground water flow. The purpose of this section, however, is geared towards the presentation of various modeling techniques, both numerical and physical, carried out with respect to the study of salt water intrusion. These studies have been categorized under two subsections, namely, sea water intrusion and noncoastal intrusion.

Models: An Enumeration

There are two types of modeling techniques primarily used in ground water flow simulation; namely, numerical and physical methods. Numerical methods are used in cases where the partial differential equations governing flow through porous media cannot be solved exactly. Various techniques have been developed for obtaining numerical solutions. The method of finite differences is one such technique (Christensen and Evans, 1974). The first step is to replace the differential equations by algebraic finite difference equations. These difference equations describe relationships among values of the dependent variable at neighboring points of the applicable coordinate space. The resulting series of simultaneous equations is solved numerically and gives values of the dependent variables at a predetermined number of discrete or "grid" points throughout the region of investigation.

The finite element technique employs a functional associated with the partial differential equation, as opposed to the finite difference method which is based on a finite difference analog of the partial differential equation (Christensen and Evans, 1974). A correspondence which assigns a real number to each function or curve belonging to some class is termed a functional. The calculus of variations is employed to minimize the partial differential equation under consideration. This is done by satisfying a set of associated equations called the Euler equations. Therefore, a functional for which the governing equations are the Euler equations is sought, and then the minimization problem is solved directly, rather than solving the differential equation. The procedure is continued by partitioning the flow field into elements, formulating the variational functional within each element, and taking derivatives with respect to the dependent variables at all nodes of the elements. The equations of all the elements are then collected. The boundary condition is expressed in terms of nodal values and incorporated into the equations. The equations are then solved.

A third technique for obtaining numerical solutions is the method of relaxation. This method may be applied to steady state problems which are adequately described by Laplace on Poisson equations (Christensen and Evans, 1974). The process involves obtaining steadily improved approximations of the solution of simultaneous algebraic difference equations. The first step of the procedure is to replace the continuous flow domain under investigation by a square or rectangular grid system. The governing differential equation is also replaced by corresponding difference equations. Next, a residual, say R_0, is defined corresponding to the point 0 on the grid; R_0 represents the amount by which the equation is in error at that point. If all values of the equation are correct, R_0 will be zero everywhere. In the initial computational step, values are assigned at all grid points and, in general, the initial residuals will not be zero everywhere. The process now consists in adjusting values at each point so that eventually all residuals approach zero, to at least some required accuracy. The reduction of residuals is achieved by a "relaxation pattern" which is repeated at different grid points so as to gradually spread the residuals and reduce their value.

It is often observed that direct analytical solutions are frequently inadequate or impractical for engineering application; in many instances, the analytical solutions thus obtained are difficult to interpret in a physical context. In an attempt to circumvent some of the shortcomings of a purely mathematical approach, physical models and analog methods are frequently employed.

Modeling, the technique of reproducing the behavior of a phenomenon on a different and more convenient scale, is an analog which has the same dimensions as that of the prototype, and in which every prototype element is reproduced, differing only in size. Consequently, the need for complete information concerning the flowfield of a prototype system is obvious and no method of solution can bypass these requirements. However, in many practical cases involving complicated geology and boundary conditions, it is usually sufficient to base the initial construction of the analog on available data and on rough estimates of missing data. The analog is then calibrated by reproducing in it the known past trend of the prototype. This is achieved by adjusting the various analog components until a satisfactory fit is obtained. Once the analog reproduces the past trend reliably, and within a required range of accuracy, it may be used to predict the response of the prototype.

One example of a physical model is a sandbox model, a reduced scale representation of a natural porous medium domain composed of a rigid, watertight container, a porous matrix filler (sand, glass beads, or crushed glass), one of several fluids, a fluid supply system and measuring devices (Christensen and Evans, 1974). The box geometry corresponds to that of the investigated flow domain, the most common shapes being rectangular, radial and columnar. Transparent material is preferred for the box construction, especially when more than one liquid may be present and a dye tracer is to be used. Porosity and permeability variations may be simulated by varying the corresponding properties of the material used as a porous matrix, which may be anisotropic. Piezometers and tensiometers may be inserted into the flow domain of the model for measuring piezometric heads.

Sea Water Intrusion

When ground water is pumped from aquifers that are in hydraulic connection with the sea, gradients are established that may induce a flow of salt water from the sea toward the well. This migration of salt water into fresh water aquifers is known as sea water intrusion. Both numerical and physical models can be used to simulate the transient positions of the phenomena.

Numerical Modeling

Numerical modeling of the steady interface in a confined aquifer by the use of finite element method has been reported by Cantatore and Volker (1974). In the work described, limitations of the Dupuit flow, regular geometry, homogeneity and isotropy of the medium were overcome. Finite element modeling has also been investigated by Cheng (1975). The published article deals with the development of the basic theory for the finite element method as applied to ground water flow, as well as basic schemes for application of the method.

Using a numerical model based on the finite element method, Christensen (1978) simulated the injection of fresh water through multiple wells into a thin confined or leaky saline aquifer in which there may be a pre-existing head gradient. A two-dimensional, finite element model of salt water intrusion in ground water systems has been reported by Contractor (1981). The report describes the development and use of a sharp interface model to simulate salt water intrusion into a coastal and insular ground water aquifer. The model, using linear triangular elements, can simulate steady or unsteady conditions in confined or unconfined aquifers. Sea water encroachment in a coastal aquifer, where the coupled nonlinear conservation equations of mass and of salt are formulated and solved by the finite element method with the aids of iteration and underrelaxation, has been reported by Lee and Cheng (1974).

Kashef (1975), in his report on management of retardation of salt water intrusion in coastal aquifers, used a horizontally displaced mesh technique to determine the free surfaces in rectangular earth sections or cores of earth dams by the finite element method. Kashef and Safar (1975), in their comparative study of fresh and salt water interfaces using finite element and simple approaches, found that both methods were in close agreement. Kashef and Smith (1975) have also conducted a study to find the extent of the progress of salt water intrusion due to the operation of one discharge well where various conditions of aquifer properties, pump capacities, natural flows, time aspects, and well locations were taken into account.

Pinder (1975) has reported on using the Galerkin finite element method to solve a set of nonlinear partial equations which describe the movement of the salt water front in a coastal aquifer, where pressure and velocities are obtained simultaneously in order to guarantee continuity of velocities between elements. Sa da Costa and Wilson (1979) have also reported on the development of a sea water intrusion model where the finite element method with a Galerkin formulation is used to solve the vertically-integrated governing equations. The model accounts for a variety of aquifer situations that occur in the field. Segol and Pinder (1976) used the Galerkin finite element of the salt water front in the Cutler area of the Biscayne Aquifer near Miami, Florida. Segol, Pinder, and Gray (1975) developed a computer program for solving the two-dimensional equations of ground water flow and mass transport using the Galerkin finite element method.

Tyagi (1975) conducted an investigation concerning the hydrodynamics of the intrusion and dispersion of a fresh-salt water interface in coastal ground water systems with the primary attention on developing a finite element model. A computer model was designed by Hall (1975) to simulate salt water intrusion and the effects of both fresh and salt water wells on the movement of the salt water wedge. Land (1975) investigated using a digital model for aquifer evaluation and an analytical technique for predicting the movement of the salt water front in response to a change of ground water flow into the ocean. A mathematical model of the movement of fresh and salt waters in coastal aquifers in which fresh water is withdrawn by wells has been presented by Kashef (1968). The three-dimension flow systems are developed by analogy to the two-dimensional finite-length systems.

Physical Modeling

Cahill (1967) reported on the use of physical model studies of the cyclic flow of salt water in a coastal aquifer. The physical model studies demonstrated visually that cyclic flow takes place in the denser fluid when fresh water moves seaward over intruding ocean water. These studies indicated that the degree of salt water intusion is controlled primarily by the flow of the fresh water rather than by tidal effects.

Christensen and Evans (1974) reported on the use of a Hele-Shaw model for making long-range predictions of the effects of increased demands on the Floridan aquifer. The Hele-Shaw, or viscous flow analog, is based on the similarity between the differential equations governing two-dimensional, saturated flow in a porous medium, and those describing the flow of a viscous liquid in a narrow space between two parallel planes. In the Hele-Shaw model the planes are transparent plates, and the plates are usually mounted in a vertical or horizontal orientation. All pertinent geological and hydrological features of the area were included in the study. Steady state characteristics of the aquifer system have been considered. In particular, the long-term effects due to pumping and artificial recharge were examined.

Steady state experimental studies with a viscous analog of the aquifer system in central Long Island, New York, have been conducted (Collins, 1972). The studies demonstrated a significant interaction between surface accretion, stream base flow, well recharge, and the degree of salt water intrusion.

Kashef (1970), in his report of model studies of salt water intrusion, presented a summary of the history of viscous flow model studies of salt water intrusion, with emphasis on the Hele-Shaw model. In addition to the study of the steady state cases of fresh water flow, three transient conditions were experimentally studied.

The suitability of a Hele-Shaw model for studying potable water availability on a long oceanic island has been investigated by Kritz (1972). The investigation dealt with the use of the model to predict maximum safe pumping rate, and the approximate location of the fresh water-salt water interface under various conditions of infiltration, island characteristics, and pumping. Shelat, Mehta, and Patel (1975) also used the Hele-Shaw model to investigate some aspects of subsurface salt water intrusion. Stegner (1972) has investigated using the Hele-Shaw model for the Rehoboth Bay area, Delaware. The results included the amount of fresh water being discharged to the sea and the extent of salt water intrusion under various imposed natural conditions.

Noncoastal Intrusion

Salt water intrusion into ground water can be caused by other phenomena such as leaky wells, upconing, and natural occurring salinity. Both numerical and physical models can be used for these noncoastal activities for the prediction of the transient nature of salt water intrusion.

Numerical Modeling

Chandler and McWhorter (1975) have investigated the upconing of saline water in response to pumping from an overlying layer of fresh water by numerical integration of the governing differential equation. There exists an optimum well penetration into the fresh water layer which permits maximum discharge without salt water entrainment. The optimum penetration increases as the vertical permeability is reduced relative to the horizontal permeability.

Curtis (1983) has reported on a computer model that simulates the movement of an interface separating two incompressible fluids, namely the fresh and salty ground water. The technique is a modification of the method of modeling interface movement by the use of distributed singularities originally presented by De Tosselin de Jong.

Kleinecke (1971) has reported on mathematical modeling of fresh water aquifers having salt water bottoms. A method was described for modeling the flow and salinity of water in an aquifer which has a fresh water lens overlying salt water. Reddell and Sunada (1970) derived a flow equation for a mixture of miscible fluids by combining the law of conservation of mass, Darcy's law, and an equation of state describing the pressure-volume-temperature-concentration relationship. The result is an equation involving two independent variables, pressure, and concentration.

Rofail (1977), in his report on mathematical modeling of stratified ground water flow, considered the problem of salt water intrusion. The problem was addressed generically where the hydrological boundaries can take on any shape and the impervious bed can take on any configuration. The recharge and discharge can also be considered according to any given function of time.

Physical Modeling

Ahmed (1972) has reported on an experimental model study conducted to determine the behavior of induced motion in a salt water body overlain by fresh water. The development of the saline mound to a steady state was recorded for various discharges for a given concentration.

SEA WATER INTRUSION CONTROL

Depending upon water usage, the purpose of sea water intrusion control methods may include one or more of the following (Kashef, 1977):

(1)   complete or partial prevention of the seaward movement of fresh water;

(2)   increasing the fresh-water pressures in order to increase the rate of flow within the aquifer or the size of the fresh-water lens; and

(3)   regulating by various methods the activities of water withdrawal in a certain zone in order to maintain a state of sea water intrusion that will not lead to critical water quality deterioration.

Considerable attention has been focused on methods to control sea water intrusion in order to protect local ground water supplies since as little as 2 percent of sea water in fresh water can render it unpotable (Todd, 1974). As such, several methods have been suggested to control sea water intrusion. Each of the methods is generally applicable to a particular type of situation, and the choice of method will be dependent on the problem to be solved. Newport (1977) suggested the following methods for controlling sea water intrusion:

(1)   reduce pumping;

(2)   relocate wells

    (a)   move wells inland

    (b)   disperse wells to eliminate areas of intense pumping;

(3)   directly recharge aquifer;

(4)   fresh water recharge into wells paralleling the coast, forming a hydraulic barrier; and

(5)   create a trough parallel to the coast by evacuating encroaching salt water from wells.

Todd (1974) adds the following techniques:

(6)   extracting sea water before it reaches the fresh water pumping zone;

(7)   extraction/injection combination; and

(7)   constructing impermeable subsurface barriers.

The first two listed methods can be combined into 'Pumping Pattern' control, and the next three into 'Artificial Recharge' techniques. The sixth method produces an 'Extraction Barrier', while 'Extraction/Injection Combination' involves artificial recharge and an extraction barrier. Finally, 'Subsurface Barriers' can be used to control sea water intrusion.

Pumping Patterns

Sea water is 2.5 percent heavier than fresh water. Therefore, a 41-foot column of fresh water is required to balance a 40-foot column of sea water. Based on this relation, within a reasonable distance from the ocean, each foot of fresh water above sea level theoretically indicates the presence of 40 feet of fresh water in the aquifer below sea level. The constant threat of sea water moving inland requires that fresh-water levels remain or be maintained as high as practicable above sea level (Klein et al., 1975).

Based on the difference in density between sea water and fresh water, proper pumping rates can maintain the fresh water at a desirable piezometric head. This can be accomplished in one of two ways: (1) reducing the pumping rate so that the fresh water piezometric head does not significantly decline; or (2) positioning the fresh water wells far enough inland so that the cone of depression does not fall below sea level. Figure 10 illustrates how a proper hydraulic gradient can be maintained by reducing the pumping rate (Todd, 1974). Figure 11 indicates the benefits of moving pumping wells further inland (U. S. Environmental Protection Agency, 1973).

A variation of the relocation of pumping wells has been suggested for the sea water intrusion which occurred in Glynn County, Georgia. Gregg (1971) described the situation as an area where three industrial users had depleted the ground water to such an extent that sea water intrusion had occurred. The industries were the major users of the 130 million gallon per day withdrawal.

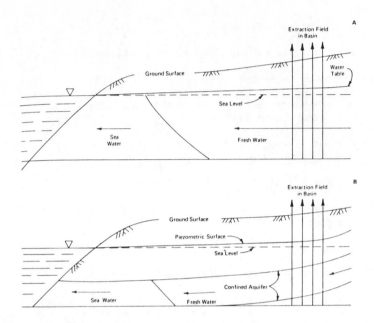

Figure 10:   Hydrologic Conditions Showing Partial Recovery from Intrusion
with a Reduction in Pumping Rate:

A--In an Unconfined Ground Water Basin
B--In a Confined Ground Water Basin

Each had their own centralized well field, and a potentiometric map of the area
showed that large cones of depression were centered under these three well
fields. The obvious solution, according to Gregg (1971), was to decentralize the
existing pumping patterns.   With the well fields dispersed, the cone of
depression would cover a larger areal extent, but the fresh water level would be
maintained at a higher elevation.

Although the decentralization of the well fields was found to be too
expensive (the actual solution was to pump sea water out of the aquifer before
it reached the well fields), the suggestion provides an excellent opportunity for
coastal communities planning to develop a fresh water well field. By initially
spreading the well network over a large area, future sea water intrusion
problems may be reduced or eliminated.

The Department of Water Resources of The Resources Agency of the
State of California (1975) makes two other practical suggestions for controlling
sea water intrusion with correct pumping patterns. First, water conservation is
a method that is applicable to all basins. Reducing demand reduces the amount
of water that must be pumped from a ground water basin.   Successful
application of this method depends on convincing all water users in an area of
the urgent need to use less water. Secondly, limitation of ground water
extractions is a method that is applicable to basins in which other sources of

water could be developed. Alternative sources could include imported water, locally-developed surface or ground water, and reclaimed waste water.

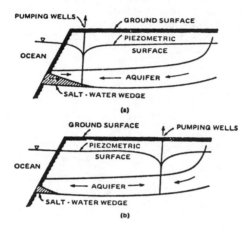

Figure 11:   Control of Sea Water Intrusion in a Confined Aquifer by Shifting Pumping Wells from (a) Near the Coast to (b) An Inland Location (U.S. Environmental Protection Agency, 1973)

The cities of Clearwater and Dunedin, Florida, are faced with the possibility of having to abandon the local aquifer as a source of 50 percent of their supply because of increases in salinity in their production wells. One suggestion for alleviating the problem is to modify current ground water pumping patterns. Cherry (1980) described a project conducted to determine the most efficient arrangement of wells and rates of pumping to obtain the greatest amount of water without causing further salt water intrusion. The project consisted of:

(1)   conducting pumping tests to determine the aquifer characteristics;

(2)   test drilling in the area to determine the location of the salt-fresh water interface; and

(3)   sampling existing wells to determine the extent of salt water encroachment.

Artificial Recharge

Sea water intrusion can be controlled by artificially recharging an intruded aquifer by the use of surface spreading areas or recharge wells. By offsetting potential overdrafts, water levels and gradients can be properly maintained. Spreading areas are most suitable for recharging unconfined aquifers, and recharge wells are pertinent for confined aquifers (U.S. Environmental Protection Agency, 1973). Artificial recharge produces a hydraulic barrier which has the effect of a physical barrier to sea water intrusion. These hydraulic barriers are created with either recharge wells or elongated surface recharge zones parallel and quite close to the shoreline.

An elongated surface recharge zone creates what is known as "fresh water ridge". Maintenance of a fresh water ridge near the shoreline creates a hydraulic barrier by raising the piezometric surface above sea level. The enhanced piezometric head prevents sea water from moving inland. Figure 12 depicts the use of a fresh water ridge to prevent sea water intrusion into an unconfined aquifer (Todd, 1974).

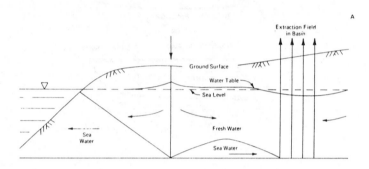

Figure 12:   Use of Surface Spreading to Create a Fresh Water Ridge to Control Sea Water Intrusion

Another type of fresh water ridge can be achieved with a line of recharge or injection wells paralleling the coast. This technique is useful with both unconfined and confined aquifers. Injection of fresh water into an unconfined aquifer will have the same effect as surface recharge. Injection into a confined aquifer, as shown in Figure 13, also prevents sea water intrusion by raising the piezometric head of the fresh water aquifer (Todd, 1974). A fresh water ridge control system based on injection wells is in operation in Los Angeles County, California (Kashef, 1977). Before the project started, it was estimated that 1,200 million m^3 (1 million acre-ft) of water was withdrawn from storage, lowering the water levels more than 31 m (100 ft) below sea level. The present total annual water requirement is estimated at 719 million m^3. This demand is partially met by importing 376 million m^3 for direct use and withdrawing the balance from the aquifers. The principal capital cost of the freshwater barrier was $15 million (excluding costs of water-rights litigation). The annual costs were $4.4 million for administration, water master costs, operation and maintenance of spreading grounds and injection wells, and the purchase of supplemental water supplies. The last item alone cost $3.3 million (Kashef, 1977).

The Los Angeles County system uses water from winter storms as the source of injection water. Bruington (1968) describes the system as a line of injection wells which creates a pressure field much like a continuous curtain wall. Waterlogging was controlled in low areas by seaward extraction along with pressure-ridge injection. Injected water was of high quality to prevent clogging of wells, increase well life, and reduce well-cleaning costs.

Observation wells were needed at critical points along the barrier line to monitor the ground water elevation. The location of the barrier line allowed a certain amount of salt water to be cut off, or trapped behind the barrier.

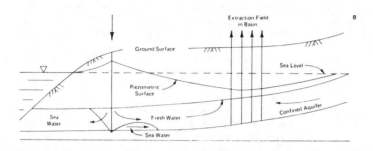

Figure 13: Use of Injection Wells to Create a Fresh Water Ridge to Control Sea Water Intrusion

Another sea water intrusion control system using injection wells occurs in northern California (Jeffers, 1982). The Santa Clara Valley Water District owns and operates a water reclamation facility located in the Palo Alto Baylands area. The purpose of the facility is to provide reclaimed water suitable for injection into the ground water, thereby providing a salt water intrusion barrier and, secondarily, to provide a research facility for various ongoing projects. The NASA Ames Research Center's Water Monitor System is to collect data on the water reclamation process train. The Water Monitor System is a continuous on-line water quality monitoring system which automatically measures 14 water quality parameters in addition to 9 trace halocarbons (Jeffers, 1982).

In Orange County, California, a sea water intrusion control program has been underway since 1965. The Orange County Water District (OCWD) and the Los Angeles County Flood Control District (LACFCD) joined in the effort of creating the Alamitos Barrier and the Talbert Barrier (Orange County Water District, 1983). The barriers are generated by injecting reclaimed wastewater into the fresh water aquifer.

The injection rate into the entire Alamitos Barrier is approximately 4,000 acre-feet a year, of which less than 20 percent is in the Orange County Water District. During the mid-1970's it was suspected that intrusion was continuing in two or three areas in Orange and Los Angeles Counties. One of these was just west of the San Gabriel River in the middle of the Alamitos Barrier, and an additional injection well was constructed and put into operation in March, 1978. Another area of intrusion was suspected around the east end of the Alamitos Barrier in the OCWD area, and during the summer of 1982 the District constructed two additional injection wells and two observation wells at a cost of approximately $400,000. The two new injection wells were put into operation in October, 1982, and they recharge water at a rate of approximately 1.4 cfs. Within one month the two new injection wells raised the groundwater levels in the immediate area approximately four feet, a good indication that the wells are having a positive effect.

The Talbert Barrier has been in operation since 1976, injecting water which, in part, has been produced from wastewater recovery. The barrier was effective in halting further migration of seawater intrusion during the drought years 1976-77. With the help of the wet years 1978-80, the intrusion of brackish water was pushed further seaward. Currently, the groundwater levels in the Orange County basin are relatively high, and the amount injected at the Talbert Barrier is low. During calendar year 1982, only 7,219 acre-feet of water was injected in the Talbert Barrier (Orange County Water District, 1983).

The problem of sea water intrusion on Long Island, New York, was discussed by Montanari and Waterman (1968). Although only the perimeter of Long Island is affected, the problem is compounded by direct contact between the aquifers and the Atlantic Ocean. All Long Island ground water comes from infiltration with approximately 50 percent absorption into the aquifers. Since 1933, legislation has been enacted to regulate ground water extraction and to license well drillers. The main causes of salt water intrusion are overpumpage and a reduction in the availability of recharge water as a result of urbanization. Projects undertaken to relieve overpumpage were: (1) reservoirs to collect surface water; (2) acquiring pumping rights of sparsely-populated areas; and (3) construction of a desalination plant. Current activities include the use of treated wastewater for injection.

Extraction Barrier

Reversing the ridge method, a line of wells may be constructed adjacent to and paralleling the coast and pumped to form a trough in the ground water level. By having these discharge wells penetrate the sea water wedge and extract the sea water, a fresh water aquifer can be protected. The extraction barrier creates a stationary wedge where the sea water's piezometric head is reduced and maintained at a lower level than the fresh water's piezometric head. Figure 14 illustrates an extraction barrier for both unconfined and confined aquifers (Todd, 1974). As long as the sea water's piezometric head is lower than the fresh water's piezometric head, sea water is prevented from intruding inland.

The extraction barrier can be an effective control measure and, as in Glynn County, Georgia (Gregg, 1971), can be the only required method. The primary disadvantages are, however, a reduction in the usable storage capacity of the basin, the expense of drilling, operating, and maintaining the extraction wells, and the waste of the fresh water that is necessarily removed along with the sea water.

Extraction/Injection Combination

By combining an extraction trough and a fresh water ridge, a more effective barrier can be created. Generating an extraction trough near the shore line and a fresh water ridge more inland allows a significant difference in the sea water and fresh water piezometric heads. The difference in piezometric heads creates the barrier to sea water intrusion. An illustration of the use of this extraction/injection combination control technique can be seen in Figure 15 (Todd, 1974).

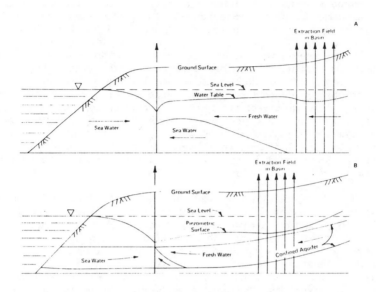

Figure 14:   Use of an Extraction Barrier to Control Sea Water Intrusion

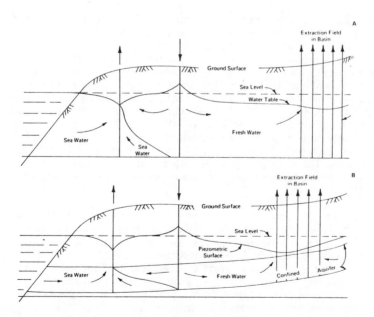

Figure 15:   Use of an Extraction/Injection Combination Barrier to Control
Sea Water Intrusion

An example of the use of the extraction/injection combination can be found in Palo Alto, California. Sheahan (1977) describes a sea water intrusion barrier consisting of a series of injection wells used to inject 2.0 million gallons per day (7.6 million liters per day) of reclaimed wastewater into a shallow aquifer. The injected water is subsequently removed by a similar system of extraction wells to avoid any possible degradation of the water supply aquifers from this source, and to allow reuse of the reclaimed wastewater. The investigation phase included test drilling, aquifer testing, and injection testing to determine the feasibility of the extraction/injection concept. The location and spacing of the trough and ridge were optimized by using a digital computer model. Double-cased, double-screened wells were constructed using corrosion-resistant materials, and the wells were designed for ease of routine maintenance. Injection and extraction are computer controlled by sensing piezometric levels in a series of monitoring wells.

Kashef (1977) notes that an extraction/injection combination would result in the waste of substantial amounts of fresh water. Another disadvantage would be the high cost of operating two systems, as well as the requirement for many more wells than when using either technique individually. The advantage of using the combination is that both extraction rates and recharge rates could be reduced as compared to either individual system (U.S. Environmental Protection Agency, 1973). Furthermore, in cases where an extraction system produces a real danger of land subsidence, recharge wells must be used (Kashef, 1977).

Subsurface Barriers

Subsurface barriers are physical barriers to sea water intrusion, as opposed to the hydraulic barriers in the control methods mentioned above. The subsurface barrier is an impermeable, vertical wall which is placed inland to restrict the movement of sea water. Barriers usually take one of three different forms: slurry walls, grout cutoffs, or steel sheet piles. A study undertaken to evaluate the effectiveness of subsurface barriers (Knox, 1983) provided descriptions of the three different forms. Slurry wall construction involves pumping a slurry made of water and bentonite clay into a trench while excavation of the trench is proceeding. The slurry maintains wall stability through hydrostatic pressure, thus decreasing the required width of excavation. As excavation proceeds, backfilling with a soil-bentonite or cement-bentonite mixture follows. The backfill combines with the slurry to form an impermeable membrane. The advantages and disadvantages are listed in Table 5 (Knox, 1983).

Grout cutoffs are another type of impermeable barrier and are constructed by injecting a liquid, slurry, or emulsion under pressure into the soil. The fluid injected will move away from the point of injection to occupy the available pore spaces. As time passes, the injected fluid will solidify, thus resulting in a decrease in the original soil permeability and an increase in the soil-bearing capacity. Like slurry walls, grouting techniques have been used for years in the construction industry, but are only now finding applications for ground water pollution control. Their advantages and disadvantages are described in Table 6 (Knox, 1983).

Sheet piling involves driving lengths of steel that connect together into the ground to form a thin impermeable permanent barrier to flow. Sheet piling

Table 5:   Advantages and Disadvantages of Slurry Walls (Knox, 1983)

Advantages	Disadvantages
1. Construction methods are simple.	1. Cost of shipping bentonite from west.
2. Adjacent areas not affected by ground water drawdown.	2. Some construction procedures are patented and will require a license.
3. Bentonite (mineral) will deteriorate with age.	3. In rocky ground, over-excavation is necessary because of boulders.
4. Leachate-resistant benton-ites are available.	4. Bentonite deteriorates when exposed to high ionic strength leachates.
5. Low maintenance require-ments.	
6. Eliminate risks due to strikes, pump breakdowns, or power failures.	
7. Eliminate headers and other above ground obstructions.	

materials also include timber and concrete; however, their application to polluted ground water cases is in doubt due to corrosive actions and costs, respectively. Sheet piling is also a relatively old technology, just now finding applications for ground water pollution control.   The advantages and disadvantages of steel sheet piles are given in Table 7 (Knox, 1983).

All three technologies share a common factor which may limit the use of subsurface barriers for controlling sea water intrusion: they must be adequately connected to an underlying impermeable geologic zone. If a continuous connection of the barrier to the impermeable formation cannot be assured, there will exist a window or passage for sea water to move through.

Figure 16 illustrates how a subsurface barrier would prevent sea water intrusion (Todd, 1974). Knox (1983) points out that three criteria must be met for subsurface barriers to be effective. First, the in-place permeability of the barrier must be very small and should remain so over time.   Second, an impermeable formation to which the barrier can be adequately attached must exist at a reasonable depth.   Third, an adequate connection between the underlying impermeable formation and the barrier must be constructed. Barriers are more effective in soils that are more conductive horizontally than vertically; for example, stratified soils. The most important result of this analysis was that in the long run a partially penetrating barrier is actually less effective in retarding sea water intrusion than no barrier.   The effect of

Table 6:   Advantages and Disadvantages of Grout Systems (Knox, 1983)

Advantages	Disadvantages
1. When designed on basis of thorough preliminary investigations, grouts can be very successful.	1. Grouting limited to granular types of soils that have a pore size large enough to accept grout fluids under pressure yet small enough to prevent significant pollutant migration before implementation of grout program.
2. Grouts have been used for over 100 years in construction and soil stabilization projects.	2. Grouting in a highly layered soil profile may result in incomplete formation of a grout envelope.
3. Many kinds of grout to suit a wide range of soil types are available.	3. Presence of high water table and rapidly flowing ground water limits groutability through a. extensive transport of contaminants b. rapid dilution of grouts
	4. Some grouting techniques are propietary.
	5. Procedure requires careful planning and pretesting. Methods of ensuring that all voids in the wall have been effectively grouted are not readily available.

partially penetrating barriers is to provide an initial delay in sea water movement, but then actually accelerate the movement of the sea water out under the barrier.

NONCOASTAL SALT-WATER INTRUSION CONTROL

Saline ground water can be found at some depth throughout the contiguous United States. In general, however, saline groundwaters occur in only a few geologic settings which can be divided into: sedimentary basins; structural troughs; thermal areas; and miscellaneous isolated occurrences. Since one or more of these geologic settings are common in almost every state in the nation, it is not unreasonable to assume that salt-water intrusion has occurred, or may potentially occur, in isolated or regional circumstances in every state. Since

Table 7:   Advantages and Disadvantages of Steel Sheet Piles (Knox, 1983)

Advantages	Disadvantages
1. Construction is not diffi-cult; no excavation is necessary.	1. The steel sheet piling ini-tially is not watertight.
2. Contractors, equipment, and materials are available throughout the United States.	2. Driving piles through ground containing boulders is difficult.
3. Construction can be economical.	3. Certain chemicals may attack the steel.
4. No maintenance required after construction.	
5. Steel can be coated for protection from corrosion to extend its service life.	

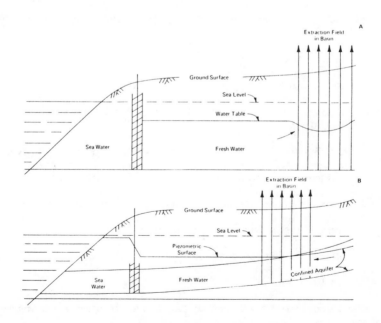

Figure 16:   Use of a Subsurface Barrier to Control Sea Water Intrusion

salt-water intrusion is or can be such a common problem, a good deal of attention has been given to control methods. Newport (1977) has suggested the following methods:

(1)   reduce pumping;

(2)   relocate wells

    (a)   move wells away from saline aquifer (horizontally or vertically)

    (b)   disperse wells to eliminate areas of intense pumping;

(3)   directly recharge aquifer;

(4)   drill scavenger wells to evacuate salt water, thus reducing the pressure on the salt water zone;

(5)   regulate subsurface disposal of oilfield brines

    (a)   select proper receptive formations

    (b)   use sound engineering techniques

    (c)   locate and properly plug abandoned wells; and

(6)   locate and plug faulty wells.

Other salt-water intrusion control methods which have been suggested include:

(7)   air injection (Roberts, 1967);

(8)   use of saline water as a resource;

(9)   impoundment and/or exportation of saline water (Holburt, 1972; and Rought, 1984);

(10)   the "Bubble Concept"--arranging control efforts so that overall salt concentration is a desired level instead of attempting to meet uniform emissions at discrete sources (Reetz, 1984);

(11)   diversion of surface water around infiltration areas (Rought, 1984);

(12)   importation of fresh water (Rought, 1984);

(13)   deep well injection-disposal (Riding, 1984); and

(14)   desalination.

These 14 control methods can be categorized into the following types of control measures:

(1)   pumping patterns;

(2)   piezometric equalization;

(3)   air injection

(4)   sealing wells; and

(5)   management techniques and other methods.

Pumping Patterns

Proper control of fresh water withdrawal can, as in sea water intrusion control, prevent non-coastal intrusion of saline water. As mentioned in Chapter 2, excessive pumping from a fresh water aquifer which overlies saline water can cause upconing of the salt water into the fresh water. Lateral movement of salt water can also be caused by excessive pumping through changing the piezometric head of the fresh water aquifer.

Controlling salt water intrusion with adjusted pumping patterns involves determining the optimum location, depth, spacing, pumping rate, and pumping sequence of withdrawal wells (U.S. Environmental Protection Agency, 1973). The objective involves maximizing the production of fresh ground water while minimizing the undermixing of fresh and saline waters. This trade-off may involve actually terminating pumping of fresh ground water, reducing the pumping rate of fresh water wells, or possibly, decentralizing withdrawal wells. The importance of controlling pumping patterns can be seen in Figure 17 (U.S. Environmental Protection Agency, 1973). Notice the nearly identical patterns in fresh water withdrawal volume and chloride concentration.

Piezometric Equalization

The relationship between fresh ground water and saline ground water can be thought of as existing in dynamic equilibrium before human activities intervene. The equilibrium occurs because the piezometric head of both waters are equal. As fresh ground water is withdrawn, its piezometric head is reduced, and the saline water tends to move towards the fresh water in an attempt to reestablish equilibrium. Since the saline water moves into the fresh water as a result of differences in piezometric heads, the U.S. Environmental Protection Agency (1973) has suggested reducing the piezometric head of the saline aquifer. This can be accomplished by drilling and pumping a well perforated only in the saline water portion of the aquifer. Although the water pumped is saline and may present a disposal problem, this method does permit the continued utilization of the underground fresh water resources without increasing intrusion.

Air Injection

The injection of air into geologic formations has been suggested as a method to control the movement of underground water (Roberts, 1967). The following description is basic to this method.

The yield from an "air-locked" well is greatly reduced by air bubbles that occupy the intergranular spaces and are held there, even against high

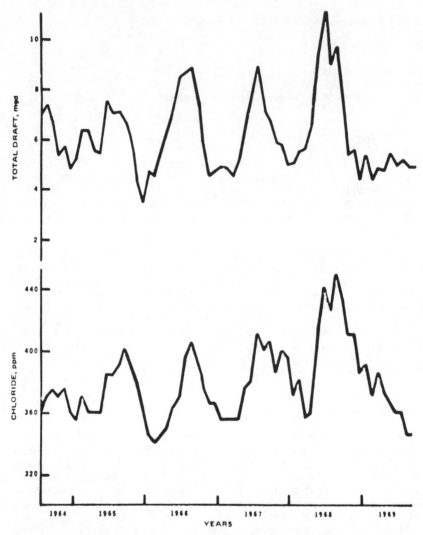

Figure 17:   Monthly Variation in Total Fresh Water Withdrawal and
Corresponding Chloride Concentration, Honolulu, Hawaii
(U.S. Environmental Protection Agency, 1973)

hydrostatic heads, by a combination of several physical forces (Roberts, 1967).
Adhesion and surface tension probably are two of the most important forces.
Observation of these low-yield conditions led to a consideration of the possible
benefits that might be obtained from the controlled injection of air into an
aquifer or other saturated mass of pervious granular material. It appears that
the injection of air could be utilized to either retard or accelerate the
movement of ground water. An air barrier, therefore, could be used to prevent
or retard the flow of salt water into a fresh water aquifer. The air barrier
would be more economical than establishing a fresh-water barrier by pumping
potable water to injection wells. Even where a fresh-water barrier proved to be
necessary, the percentage of injected potable water lost could be reduced by

the establishment of an air barrier, which would reduce the permeability of the aquifer on both sides of the fresh-water barrier.

Permeability is reduced by the presence of air in intergranular spaces. Figure 18 (Roberts, 1967) represents the exponential relationship between relative permeability and percent liquid saturation. As the percentage of liquid saturation goes down (that is, an increase in intergranular air), the relative permeability also goes down. An effective air barrier to salt water intrusion would require first displacing the water in a zone ahead of the salt water interface with injected air, and secondly replenishing that air as it migrates or becomes absorbed in the water.

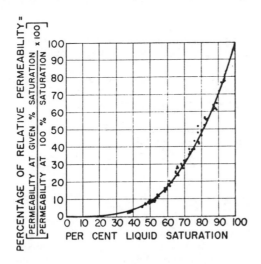

Figure 18:   A Typical Relative Permeability Curve (Roberts, 1967)

An air barrier would act somewhat like a grout curtain, although more effectively, because of the following:

(1)   Test results, such as those summarized in Figure 18, are obtained from uniform materials seldom encountered in nature; therefore, air would first be distributed by displacing the water from the more pervious zones, rapidly reducing the relative permeability, and then by infiltrating the zones of lesser permeability;

(2)   A double line of air injection wells would normally be advisable; and

(3)   Air would, when injected under a continuing or repeated proper pressure, seek out and become lodged in the intergranular spaces which are often rendered inaccessible during initial grout injections.

The following considerations are basic to the development of an effective air barrier (Roberts, 1967):

(1)   Increasing the distance between the injection wells of a line would increase the economic attractiveness of the proposed barrier; the

horizontal dispersion of air from the injection well is, therefore, of primary importance; under most conditions, the horizontal dispersion might be best increased by releasing slugs of air from portable receivers; the distance that the air would travel depends upon several factors, among which size of open pores, permeability and opposing hydrostatic head are perhaps the most important;

(2)   The injected air should be prevented from leaking back up along the casing; once a blowout occurred, the injection well would be ruined;

(3)   The spacing of the wells in the injection line, and the distance between the two injection lines, would vary with the subsurface conditions;

(4)   Fairly accurate logs of the injection wells would be necessary to determine the proper depths at which air should be injected;

(5)   A monitoring system would have to be developed to provide information indicating the economic limits of injecting air and when recharging would be desirable; and

(6)   Under certain conditions, a rising blanket of air could, in effect, cause granular material to become "quick" and slide on very flat slopes.

Sealing Wells

A leaking well allows the entry of surface water into fresh water aquifers and may allow the migration of saline water into a fresh water aquifer. To minimize the movement of saline water in leaking wells, abandoned wells, and test holes, the wells should be sealed. Sealing typically involves some method of plugging the well hole. Today, the most common type of plugging material is a cement slurry, although other types of materials are also used.

Herndon and Smith (1984) describe three basic factors which influence the condition and effectiveness of a cement plug to seal a well:

(1)   the condition of the mud or drilling fluid in the hole;

(2)   the volume and type of cement used; and

(3)   placement techniques for the plug.

The first thing to be considered in setting a plug is the condition of the well and the fluid in the hole at the time of placement. It is extremely important that the well be perfectly static when the plug is placed. If there is any movement of the cement after it is placed, proper setting is unlikely. If gas is coming through a section and percolating up through the slurry, the cement could become permeable or honeycombed from gas cutting and may never seal properly. Salt water flows will do the same thing. A salt water flow moving the slurry may limit proper setting or may cause complete inhibition. A tipoff to this situation would be an increase or decrease in the mud weight (Herndon and Smith, 1984).

If wells are drilled under adverse conditions, the mud is sometimes highly treated with chemicals. Most mud treating chemicals used today are organic in nature, and many will retard the setting of cement slurries. Muds adverse to cement are primarily those that contain filtration control agents and those that contain thinners. Contamination of cement with these types of muds will seriously affect the compressive strength and cause non-setting of the cement slurry if in sufficient proportion.

The cement used should be selected to match the formation encountered; for example, salt-saturated cement should be used for plugs set in salt formations. For most plugging situations, a densified (low water ratio) cement should be used for optimum strength and maximum resistance to mud contamination. The advantages of using this type slurry are (Herndon and Smith, 1984):

(1) it is less subject to mud contamination and resultant loss of strength;

(2) it sets and gains strength more rapidly, thus reducing cement waiting time;

(3) with dispersants, such slurries have water loss control and minimize dehydration of the cement slurry which could allow fluid movement in the borehole; and

(4) the addition of salt can give improved bonding and minimize damage to shales or bentonite sands.

Plugs set by oil companies in dry holes or abandoned producing wells generally tend to follow the rules and regulations of each state's oil and gas regulatory body. Figure 19 shows typical plugging requirements set by most states (Herndon and Smith, 1984). Where rules do not exist, some companies have their own plugging procedures. Typically, for a dry hole, this might include a cement plug at the bottom of the hole, extending 50 to 100 ft above any possible producing zone, a plug at the bottom of an intermediate string and the surface pipe, and a plug at ground surface.

Management Techniques and Other Methods

Various suggestions (Baer, 1972; Holburt, 1972; Reetz, 1984; and Rought, 1984) have been given to alleviate saline ground water problems without actually preventing intrusion into fresh water. One suggestion is that fresh water could be added to natural streamflows during low flows to dilute the brine at the salt source areas. The large quantity of fresh water needed for effective dilution tends to make the concept impractical.

One recommendation is to consider saline water as a resource. By collecting brine, various minerals can be extracted. In the United States, sodium chloride (salt), sodium carbonate, sodium sulfate, potassium chloride (potash), calcium chloride, and magnesium (the ions common in ground water) are produced from brines. Brines can also yield iodine, bromine, lithium chloride, lithium carbonate, boric acid, boron, and borates. Other minerals that might be produced include potassium sulfate, cesium, rubidium, silver, zinc, lead, copper, molybdenum, and strontium.

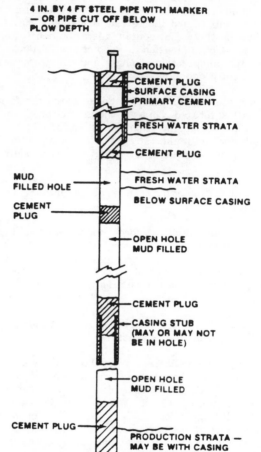

4 IN. BY 4 FT STEEL PIPE WITH MARKER
— OR PIPE CUT OFF BELOW
PLOW DEPTH

GROUND
CEMENT PLUG
SURFACE CASING
PRIMARY CEMENT

FRESH WATER STRATA

CEMENT PLUG

MUD
FILLED HOLE   FRESH WATER STRATA

                               BELOW SURFACE CASING
CEMENT
PLUG

                               OPEN HOLE
                               MUD FILLED

                               CEMENT PLUG

                               CASING STUB
                               (MAY OR MAY NOT
                               BE IN HOLE)

                               OPEN HOLE
                               MUD FILLED

CEMENT PLUG   PRODUCTION STRATA —
                               MAY BE WITH CASING
                               AND PERFORATIONS

Figure 19:   Schematic of Typical Plugs Required by Most States (Herndon and
              Smith, 1984)

Another suggestion similiar to diversion of fresh water around salt sources involves precluding infiltration from canals and other surface water channels. Introduction of water into highly saline soils and aquifers results in excessive salinity contributions. Consequently, by reducing these ground water inflows by a program of canal and lateral linings, significant reductions in salt contamination could be achieved.

WATER DESALINATION

Once salt water has intruded into an area, it may be more technically or economically feasible to desalinate the water than to turn to alternative sources or control the intrusion. One of the major problems with this process is the disposal of the waste brine. Though there are quite a few ways to desalinate water, distillation, electrodialysis, and reverse osmosis are the methods most noted.

Distillation

There are many methods of distillation, including multiple-effect evaporation, vapor-compression distillation, long-tube vertical distillation, and evaporation by solar stills or immiscible-liquid heat transfer. A two-stage vapor compression plant for desalting highly saline ground water has been built at Roswell, New Mexico, under the Federal saline-water conversion program. This plant is designed to produce 1,000,000 gal per day (700 gpm) of fresh water. The well water being treated contains more than 24,000 ppm dissolved solids, with hardness of over 3,000 ppm. Process problems have been related to pre-treatment of the feed water to avoid scale formation in the plant (Johnson Division, 1975). Formation of scale on heated metal surfaces is one of the major problems and sources of cost in distillation units. This has led to the testing of liquid-to-liquid heat transfer wherein the salt water is heated by direct contact with a hot immiscible liquid (Johnson Division, 1975).

Electrodialysis

The electrodialysis process takes place in a chamber that has been divided into three compartments by two ion sensitive membranes. One membrane allows only negatively charged ions to pass through; the other allows only positively charged ions to pass. When an electric current is passed through the water, the ions collect in the outer chambers close to the electrodes, and the water in the middle chamber is demineralized to a considerable degree. Webster, South Dakota, uses this method for treating its municipal water supply. The installation was set up as a demonstration facility in 1961 by the Office of Saline Water of the U.S. Department of the Interior. It was originally designed to produce 250,000 gallons per day, but its capacity has been increased to 320,000 gallons per day (Johnson Division, 1975). Other electrodialysis units serving municipal water systems include those at Coalinga, California, and Buckeye, Arizona.

Reverse Osmosis

Reverse osmosis is the newest and perhaps the simplest method for desalting water. In osmosis, if salt water and pure water are placed on the opposite sides of a semipermeable membrane, pure water slowly passes through the membrane into the salt water side without any outside force needed. If pressure is placed on the salt water side, this process reverses, and the water from that side passes into the pure water side, concentrating the salts on the salt side. This is the basic principle of the reverse osmosis process. In 1973 a pilot unit was established at Coalinga, California, under the supervision of the University of California, that produced 5,000 gallons per day. After a year of successful operations, it was enlarged (Johnson Division, 1975).

The largest plant in operation in the United States at this time is Water Factory 21, located in Fountain Valley, Orange County, California. This plant, which became operational in 1976, is capable of desalting five mgd by reverse osmosis. In October, 1981, the Orange County Water District completed a three-year program on evaluation of membrane processes in wastewater reclamation. One of the conclusions was that reverse osmosis is effective not only in desalination, but also in the removal of heavy metals, organics, and viruses. Also, new experimental low pressure, high flux membrane tests indicate greater salt rejection is possible with less energy consumption (Orange County Water District, 1983).

Another reverse osmosis plant is under construction at Yuma, Arizona, and should be on line in 1988. This plant is being built by the U.S. Bureau of Reclamation to honor the United States' agreement with Mexico to deliver to them approximately 1.4 million acre-ft. of water annually with an average salinity of less than 115 milligrams per liter. The plant is sized to produce 72 mgd (Trompter and Suemoto, 1984).

INSTITUTIONAL AND LEGAL ASPECTS

There are several institutional and legal considerations associated with salt water intrusion control, including water rights, federal policies and laws, and potential management organizations such as water management districts.

Water Rights

Any attempt to control an activity involving the use of ground water will probably involve, and usually be in conflict with, vested water rights. Because states have adopted ground water laws much more recently than surface water laws, ground water laws tend to follow the statutes already adopted for surface water and display the same philosophies. However, ground water laws have become much more complex. There are four basic ground water doctrines: (a) Absolute Ownership, (b) Reasonable Use, (c) Correlative Rights, and (d) the Doctrine of Prior Appropriation. Some states use more than one doctrine, depending on the type of ground water involved.

Absolute Ownership Doctrine

The Absolute Ownership Doctrine is the English common law version of the Riparian Doctrine, which simply states that the landowner can withdraw as much water as desired from beneath owned land without liability to neighbors in so doing. The doctrine gives absolute ownership of the ground water beneath the land to the landowner. The water is deemed to be part of the soil (Interagency Task Force, 1979). States using this doctrine include Texas, Louisiana, Arkansas, Missouri, Minnesota, Indiana, Ohio, Pennsylvania, Vermont, Massachusetts, Connecticut, Maine, New Jersey, Rhode Island, South Carolina, Georgia, Alabama, Mississippi, and parts of California.

Reasonable Use Doctrine

As more knowledge was acquired about ground water and its flow under the ground surface and how interferences caused by users affect landowners above the same aquifer, many of the states modified the Absolute Ownership Doctrine to limit the landowner to a reasonable use. This concept restricts the landowner's rights to water from those of absolute ownership to that of

Reasonable Use, recognizing equal rights of all to ground water resources (Interagency Task Force, 1979). States employing this system of law include Arizona, Nebraska, Iowa, Illinois, Michigan, Kentucky, Tennessee, Florida, North Carolina, Virginia, Delaware, West Virginia, Maryland, New York, New Hampshire, and Wisconsin.

Correlative Rights Doctrine

The State of California has gone one step further with the Reasonable Use Doctrine in formulating the Correlative Rights Doctrine. The landowner's use must not only be reasonable, but must be correlated with the use of others, particularly during times of water shortage. When supplies become insufficient because of a drought or the drawdown effect, each landowner is entitled to water in proportion to the percentage of owned land in relation to all other lands overlying the groundwater supply. The doctrine is also followed in Oklahoma and somewhat in Colorado and Nebraska. In Colorado, SB-213 allocates water in a confined aquifer based on one percent of the available in-storage supply annually (Interagency Task Force, 1979).

Doctrine of Prior Appropriation

As in surface water supplies, under the Doctrine of Prior Appropriation, ground water is considered the property of the state and the resource is subject to appropriation for beneficial use through a priority system with principles very similar to those for surface waters (Interagency Task Force, 1979). States using this doctrine are Washington, Oregon, Idaho, Nevada, Montana, Wyoming, Utah, Colorado, New Mexico, North Dakota, South Dakota, Kansas, Alaska, and, in part, California.

Federal Policies and Laws

The Federal Government has recognized the rights and responsibilities of states for the control and protection of surface and ground water. Section 8 of the Reclamation Act of 1902 (32 Stat. 388, 1902) explicitly provides that the Secretary of the Interior shall obtain water rights for reclamation projects in accordance with state water laws. The same provision, or one expressing the same intent, has been included in acts amendatory of and supplementary to the original Reclamation Act (U.S. Environmental Protection Agency, 1973).

In further support of this apparently consistent Congressional intent, it is significant that there are no federal statutes governing the allocation of water resources, surface or ground, or the administration of water rights. Although periodically bills are introduced in the U.S. Congress for those purposes, they have never passed beyond the committee stage (U.S. Environmental Protection Agency, 1973).

The President's Water Policy Statement of June, 1978, noted that the states should manage water resources, and a Presidential memorandum of July, 1978, emphasized that the states are principally responsible for protecting streams and preventing ground water depletion (U.S. General Accounting Office, 1980). The Federal Government has, however, concerned itself with water quality. In 1972 the Federal Water Pollution Control Act and

Amendments, while recognizing the primary responsibility of the states to prevent and reduce water pollution, directed the U.S. Environmental Protection Agency to develop comprehensive programs for water pollution control in cooperation with state and local government and other federal agencies.

Also, in 1977, Section 208 of the Clean Water Act (33 U.S.C.) required states to submit area-wide water quality management plans to the U.S. Environmental Protection Agency that identified and proposed solutions to water quality problems. These plans must include regulatory and nonregulatory activities and best management practices selected to meet nonpoint source pollution control needs. Like the Federal Water Pollution Control Act, the Clean Water Act is directed at the protection of surface water, but some states have chosen to also include ground water in their programs (Office of Technology Assessment, 1984).

The Rural Clean Water Program (Section 208(j) of the Clean Water Act) authorized a technical assistance and cost-sharing program with rural landowners to implement the best management practices to control nonpoint source pollution for improved water quality. This program was administered by the Secretary of Agriculture through the U.S. Soil Conservation Service, with the concurrence of the Administrator of the U.S. Environmental Protection Agency. To qualify for cost-sharing funds, the conservation practices and measures must be part of an approved section of a 208 water quality management plan (Interagency Task Force Report, 1979). The funding for this section was terminated in 1981. However, grants under Sections 106 and 205(j) are now being used to support these activities (Office of Technology Assessment, 1984).

The Federal Government has also lent technical and financial assistance to states for water development projects, such as the Central Arizona Project and the Central Valley Project (California), that are directly or indirectly related to the salt water intrusion problem.

The U.S. Geological Survey (USGS) provides data and technical assistance for ground water management through its Federal/State Cooperative Program. Under this program, the Federal Government and over 500 state and local agencies share the cost of USGS investigations and research programed in collaboration with state and local agencies. These cooperative projects are designed to provide continuing appraisals of water quantity and quality and to improve hydrological information and understanding. The program makes project results available to federal, state, and local agencies for use in developing, utilizing, conserving, and managing water and land resources. More than half of the water resources data gathered in the United States, except for stream gaging, is provided by the Cooperative Program (U.S. General Accounting Office, 1980).

Water Management Districts

Legislation at the state level has provided for the creation and operation of many types of districts for water use and management. These are mostly single purpose districts such as irrigation districts, drainage districts, ground water control districts, and water improvement districts (Interagency Task Force, 1979). The U.S. Environmental Protection Agency (1973) suggested that these districts should have not only definable hydrologic boundaries but, as much as possible, should also have a social or economic identity or common interests. The powers of the organizations vary with the duties or purposes of the group. These powers may include fixing water prices; collecting assessments; planning, constructing, and maintaining waterworks; and allocating district water. Such powers are enunciated in the enabling statutes of each state (Interagency Task Force, 1979).

Kansas, Colorado, Oklahoma, and Texas have ground water management districts that provide the means for managing the water supply beneficially on a basis other than priority. In recent years, though, states have been leaning towards multiple-use water districts. These districts respond to ground water problems by: (1) vesting control of all water resources in a single basinwide entity; (2) managing ground and surface waters as if they were a single unit; and (3) implementing water conservation programs. Where this has occurred, the joint management of surface and ground waters has helped not only with salt water intrusion but also with land subsidence and reduced surface flow (U.S. General Accounting Office, 1980).

Nebraska passed legislation that dissolved the existing ground water districts in 1982. The management responsibilities are being assumed by multipurpose natural resource districts (Interagency Task Force, 1979). States such as New Mexico and Colorado have successfully handled their problems because a single entity had basinwide authority to control water use (U.S. General Accounting Office, 1980). In Florida, in 1972, a record-breaking drought that resulted in over-drafting and salt water intrusion also resulted in the passage of legislation vesting control of virtually all the states' water in what is now the State Department of Environmental Regulation and various water management districts.

California has nine Regional Water Pollution Control Boards and several smaller water districts. The districts have no legal authority to limit ground water pumping directly, but they do have legal control of water pricing (U.S. General Accounting Office, 1980). They have facilities to replenish overdrafted aquifers and use aquifer storage space to conserve water for use during dry periods.

SELECTED REFERENCES

Ahmed, N., "Salt Pollution of Ground Water", OWRR-B-043-MO(1), June 30, 1972, Missouri Water Resources Research Center, University of Missouri at Rolla, Rolla, Missouri.

Baer, T.J., Jr., "Grand Valley Salinity Control Demonstration Project", Managing Irrigated Agriculture to Improve Water Quality, Proceedings of National Conference on Managing Irrigated Agriculture to Improve Water Quality, U.S. Environmental Protection Agency and Colorado State University, 1972, pp. 109-122.

Bruington, A.E., "The Amelioration or Prevention of Salt-Water Intrusion in Aquifers -- Experience in Los Angeles County, California", Louisiana Water Resources Research Institute Bulletin 3, October, 1968, Louisiana State University, Baton Rouge, Louisiana, pp. 153-168.

Cahill, J.M., "Hydraulic Sand-Model Study of the Cyclic Flow of Salt Water in a

Coastal Aquifer", Professional Paper 575-B, 1967, U.S. Geological Survey, Phoenix, Arizona, pp. B240-244.

Cantatore, W.P., and Volker, R.E., "Numerical Solution of the Steady Interface in a Confined Coastal Aquifer", Institute of Engineers--Australian Civil Engineering Transactions, Vol. CE 16, No. 2, 1974, pp. 115-119.

Chandler, R.L., and McWhorter, D.B., "Upconing of the Salt-Water-Fresh-Water Interface Beneath a Pumping Well', Ground Water, Vol. 13, No. 4, July-August, 1975, pp. 354-359.

Cheng, R.T., "Finite Element Modeling of Flow Through Porous Media", OWRT-C-4026(9006)(4), March, 1975, Office of Water Research and Technology, U.S. Department of the Interior, Washington, D.C.

Cherry, R.N., "Hydrology of Clearwater-Dunedin, Florida", 1980, U.S. Geological Survey, Tampa, Florida.

Christensen, B.A., "Mathematical Methods for Analysis of Fresh Water Lenses in the Coastal Zones of the Floridan Aquifer", OWRT-A-032-FLA(1), December, 1978, Office of Water Research and Technology, U.S. Department of the Interior, Washington, D.C.

Christensen, B.A., and Evans, A.J., Jr., "A Physical Model for Prediction and Control of Salt Water Intrusion in the Floridian Aquifer", Publication No. 27, January, 1974, Water Resources Research Center, University of Florida, Gainesville, Florida.

Collins, M.A., "Ground-Surface Water Interaction in the Long Island Aquifer System", Water Resources Bulletin, Vol. 8, No. 6, December, 1972, pp. 1253-1258.

Contractor, D.N., "A Two-Dimensional, Finite Element Model of Salt Water Intrusion in Groundwater Systems", Technical Report No. 26, October, 1981, Water Research Institute of the Western Pacific, Guam University, Agana, Guam.

Curtis, T.G., Jr., "Simulation of Salt Water Intrusion by Analytic Elements", Dissertation Abstracts International, Vol. 44/11-B, 1983, p. 3486.

Department of Water Resources, The Resources Agency, State of California, "Sea-Water Intrusion in California: Inventory of Coastal Ground Water Basins", Bulletin No. 63-5, October, 1975.

Gregg, D.O., "Protective Pumping to Reduce Aquifer Pollution, Glynn County, Georgia", Ground Water, Vol. 9, No. 5, September-October, 1971, pp. 21-29.

Groot, J.J., "Salinity Distribution and Ground-Water Circulation Beneath the Coastal Plain of Delaware and the Adjacent Continental Shelf", Open File Report No. 26, May, 1983, Delaware Geological Survey, Newark, Delaware.

Hall, P.L., "Computer Model Guides Salt Water Well Construction", Johnson Drillers Journal, Vol. 47, No. 1, January-February, 1975, pp. 1-2.

Herndon, J., and Smith, D.K., "Setting Downhole Plugs: A State- of-the-Art",

Ed., Fairchild, D.M., Proceedings of the First National Conference on Abandoned Wells: Problems and Solutions, Environmental and Ground Water Institute, University of Oklahoma, Norman, Oklahoma, 1984, pp. 237-250.

Holburt, M.B., "Salinity Control Needs in The Colorado River Basin", Managing Irrigated Agriculture to Improve Water Quality, Proceedings of National Conference on Managing Irrigated Agriculture to Improve Water Quality, U.S. Environmental Protection Agency and Colorado State University, 1972, pp. 19-25.

Interagency Task Force, Irrigation Water Use and Management, U.S. Government Printing Ofice, Washington, D.C., 1979.

Jeffers, E.L., et al., "Water Reclamation and Automated Water Quality Monitoring", EPA-600/2-82-043, March, 1982, U.S. Environmental Protection Agency, Municipal Environmental Research Laboratory, Cincinnati, Ohio.

Johnson Division, "Elements of Water Treatment", Ground Water and Wells, 4th Ed., Johnson Division, St. Paul, Minnesota, 1975, pp. 354-362.

Kashef, A.I., "Management and Control of Salt-Water Intrustion in Coastal Aquifers", CRC Critical Reviews in Environmental Control, Vol. 7, July, 1977, pp. 217-275.

Kashef, A.I., Management of Retardation of Salt Water Intrusion in Coastal Aquifers", OWRT-C-4061(9013)(2), July, 1975, Office of Water Research and Technology, U.S. Department of the Interior, Washington, D.C.

Kashef, A.I., "Model Studies of Salt Water Intrusion", Water Resources Bulletin, Vol. 6, No. 6, December, 1970, pp. 944-967.

Kashef, A.I., "Diffusion and Dispersion in Porous Media--Salt Water Mounds in Coastal Aquifers", Report No. 11, 1968, Water Resources Research Center, North Carolina State University, Raleigh, North Carolina.

Kashef, A.I., and Safar, M.M., "Comparative Study of Fresh-Salt Water Interfaces Using Finite Element and Simple Approaches", Water Resources Bulletin, Vol. 11, No. 4, August, 1975, pp. 651-665.

Kashef, A.I., and Smith, J.C., "Expansion of Salt-Water Zone Due to Well Discharge", Water Resources Bulletin, Vol. 11, No. 6, December 1975, pp. 1107-1120.

Klein, H., et al., "Water and the South Florida Environment", Water-Resources Investigation 24-75, August, 1975, U.S. Geological Survey, Washington, D.C.

Kleinecke, D., "Mathematical Modeling of Fresh-Water Aquifers Having Salt Water Bottoms", Report No. 71TMP-47, July, 1971, General Electric Company, Santa Barbara, California.

Knox, R.C., "Effectiveness of Impermeable Barriers for Retardation of Pollutant Migration", Ph.D. Dissertation, University of Oklahoma, Norman, Oklahoma, 1983.

Kritz, G.J., "Analog Modeling to Determine the Fresh Water Availability on the

Outer Banks of North Carolina", Report No. 64, April, 1972, North Carolina Water Resources Research Institute, North Carolina State University, Raleigh, North Carolina.

Land, L.F., "Effect of Lowering Interior Canal Stages on Salt-Water Intrusion into the Shallow Aquifer in Southeast Palm Beach County, Florida", Open File Report FL 75-74, 1975, U.S. Geological Survey, Tallahassee, Florida.

Larson, J.D., "Distribution of Saltwater in the Coastal Plain Aquifers of Virginia", Open File Report 81-1013, 1981, U.S. Geological Survey, Richmond, Virginia.

Lee, C., and Cheng, R.T., "On Seawater Encroachment in Coastal Aquifers", Water Resources Research, Vol. 10, No. 5, October, 1974, pp. 1039-1043.

Montanari, F.W., and Waterman, W.G., "Protecting Long Island Aquifers Against Salt-Water Intrusion", Bulletin 3, October, 1968, Louisiana Water Resources Research Institute, Louisiana State University, Baton Rouge, Louisiana, pp. 59-86.

Newport, B.D., "Salt Water Intrusion in the United States", EPA-600/8-77-011, July, 1977, Robert S. Kerr Environmental Research Laboratory, U.S. Environmental Protection Agency, Ada, Oklahoma.

Office of Technology Assessment, "Protecting the Nation's Ground Water from Contamination" OTA-O-233, October, 1984, U.S. Congress, Washington, D.C.

Oklahoma Water Resources Board, "Salt Water Detection in the Cimarron Terrace, Oklahoma", EPA-660/3-74-033, April, 1975, U.S. Environmental Protection Agency, Corvallis, Oregon.

Orange County Water District, "Annual Report: 50th Anniversary Edition", June, 1983, Orange County, California.

Pinder, G.F., "Numerical Simulation of Salt-Water Intrusion in Coastal Aquifers, Part 1", OWRT-C-5224(4214)(2), 1975, Office of Water Research and Technology, U.S. Department of the Interior, Washington, D.C.

Reddell, D.L., and Sunada, D.K., "Numerical Simulation of Dispersion in Groundwater Aquifers", Hydrology Paper No. 41, June, 1970, Department of Agricultural Engineering and Department of Civil Engineering, Colorado State University, Fort Collins, Colorado.

Reetz, G.R., "Salinity Bubbles and Offsets: A Concept Paper", Salinity in Watercourses and Reservoirs, Proceedings of the 1983 International Symposium on State-of-the-Art Control of Salinity, July 13-15, Salt Lake City, Utah, Ed., French, R.H., Butterworth Publishers, Boston, Massachusetts, 1984, pp. 79-91.

Riding, J.R.,"Salinity Control Benefits Through Regulation of Produced Water", Salinity in Watercourses and Reservoirs, Proceedings of the 1983 International Symposium on State-of-the-Art Control of Salinity, July 13-15, Salt Lake City, Utah, Ed., French, R.H., Butterworth Publishers, Boston, Massachusetts, 1984, pp. 367-375.

Roberts, G.D., "Use of Air to Influence Groundwater Movement", Artificial

Recharge and Management of Aquifers, Proceedings of the Symposium on Haifa, Publication No. 72, International Association of Scientific Hydrology, UNESCO, March, 1967, pp. 390-398.

Rofail, N., "Mathematical Model of Stratified Groundwater Flow", Hydrological Science Bulletin, Vol. 22, No. 4, December, 1977, pp. 503-512.

Rought, B.G., "The Southwestern Salinity Situation: The Rockies to the Mississippi River", Salinity in Watercourse and Reservoirs, Proceedings of the 1983 International Symposium on State-of-the-Art Control of Salinity, July 13-15, Salt Lake City, Utah, Ed., French, R. H., Butterworth Publishers, Boston, Massachusetts, 1984, pp. 115-124.

Sa da Costa, A.A.G., and Wilson, J.L., "A Numerical Model of Seawater Intrusion in Aquifers", NOAA-80021406, December, 1979, National Oceanic and Atmospheric Administration, Rockville, Maryland.

Segol, G., and Pinder, G.F., "Transient Simulation of Saltwater Intrusion in Southeastern Florida", Water Resources Research, Vol. 12, No. 1, February 1976, pp. 65-70.

Segol, G., Pinder, G.F., and Gray, W.G., "Numerical Simulation of Salt-Water Intrusion in Coastal Aquifers, Part 2", OWRT-C-5224(4214)(2), 1975, Office of Water Research and Technology, U.S. Department of the Interior, Washington, D.C.

Sheahan, N.T., "Injection/Extraction Well System -- A Unique Seawater Barrier", Ground Water, Vol. 15, No. 1, January-February, 1977, pp. 32-50.

Shelat, R.N., Mehta, P.R., and Patel, G.G., "Salt Water Intrusion into Porous Media", Journal of the Institution of Engineers (India), Vol. 55, Part Ph. 2, February, 1975, pp. 55-60.

Stegner, S.R., "Analog Model Study of Ground Water Flow in the Rehoboth Bay Area, Delaware", Technical Report No. 12, July, 1972, National Technical Information Service, U.S. Department of Commerce, Springfield, Virginia.

Stewart, M.T., and Gay, M.T., "Evaluation of a Ground Transient Electromagnetic Remote Sensing Method for the Deep Detection and Monitoring of Salt Water Interfaces", Publication No. 73, 1982, Water Resources Research Center, University of South Florida, Tampa, Florida.

Todd, D.K., "Salt-Water Intrusion and Its Control", Journal of American Water Works Association, Vol. 66, No. 3, March, 1974, pp. 181-187.

Trompeter, K.M., and Suemoto, S.H., "Desalting by Reverse Osmosis a Yuma Desalting Plant", in French, R.H., Ed., Salinity in Watercourses and Reservoirs, Proceedings of the 1983 International Symposium on State-of-the-Art Control of Salinity, July 13-15, 1983, Salt Lake City, Utah, Butterworth, Boston, 1984, pp. 427-454.

Tyagi, A.K., "Finite Element Modeling of Sea Water Intrusion in Coastal Ground Water Systems", Proceedings of the 2nd World Congress on Water Resources; Water for Human Needs, New Delhi, India, December 12-16, 1975, Vol. 3, 1975, International Water Resources Association, New Delhi, India, pp. 325-338.

U.S. Environmental Protection Agency, Office of Air and Water Programs, "Identification and Control of Pollution from Salt Water Intrusion", EPA-430/9-73-013, 1973.

U.S. General Accounting Office, "Report to The Congress - Ground Water Overdrafting Must be Controlled", Report CED-80-96, September 12, 1980, Washington, D.C.

Woodruff, K.D., "The Occurrence of Saline Ground Water in Delaware Aquifers", Report of Investigation No. 13, August, 1969, Delaware Geological Survey, Newark, Delaware.

CHAPTER 4

NATIONAL SALINITY SURVEY

This chapter provides a summary of the findings of an investigation of salt-water intrusion in the contiguous United States. The first section presents information gathered from earlier studies. The second section represents an overview of salt-water intrusion problems in the continental United States as found in this investigation.

PREVIOUS NATIONAL OR REGIONAL SUMMARIES

There are various generalized types of information pertinent to understanding the national extent of salt-water intrusion. Locations of major and minor aquifers represent one type of information. Figures 20 and 21 indicate the locations of aquifers which provide water for private, municipal, industrial, and agricultural purposes (Geraghty et. al., 1973). The locations indicate areas where salt-water intrusion may be of concern or where ground water quality may have already been degraded. Later sections of this chapter provide more detail on known problems.

Another type of information which helps to understand the possible extent of salt-water intrusion involves the salinity of surface waters. Figure 22 (Geraghty et al., 1973) indicates the total dissolved solids content of surface water throughout the conterminous United States. Obviously, the western half of the nation is affected more by surface salinity than is the eastern half.

The last type of information pertinent to understanding the possible extent of salt-water intrusion is the depth to saline ground water. If saline ground water is shallow and close to fresh water aquifers, the potential for salt-water intrusion increases. Figure 23 (Geraghty et. al., 1973) displays the depth to saline ground water in increments of 500 feet. Other than the nation's coastline, the Central Plains States and Mid-western States contain the majority of shallow saline ground water.

Keeping the above generalized information in mind, a review of two previous national surveys is appropriate. The first was a survey of salt-water intrusion in the United States published in 1977 by the U. S. Environmental Protection Agency (Newport, 1977). The report indicated that salt-water intrusion has resulted in degradation of subsurface fresh water aquifers in 43 states. The survey indicated the following summary of the types of salt-water intrusion problems and the number of states affected (Newport, 1977):

Number of States Affected	Type of Salt Water Intrusion
27	Lateral intrusion caused by excessive pumping

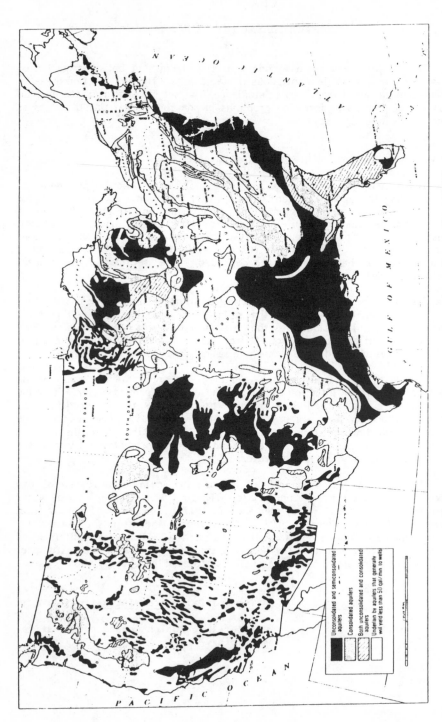

Figure 20:   Major Ground Water Aquifers (Geraghty et al., 1973)

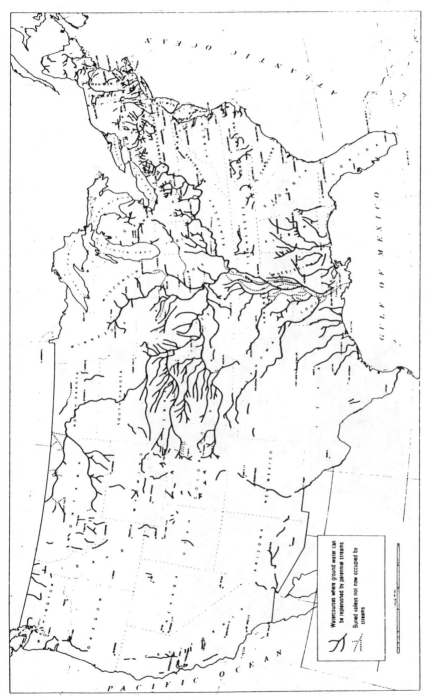

Figure 21:   Minor Ground Water Aquifers Associated with River Valleys
(Geraghty et al., 1973)

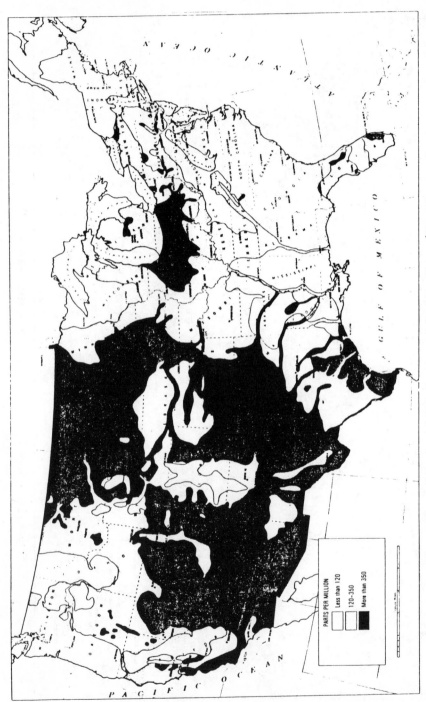

Figure 22:   Dissolved Solids Content of Surface Water (Geraghty et al., 1973)

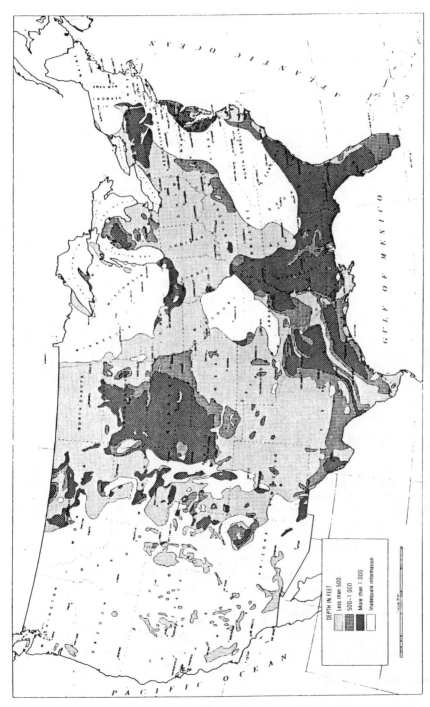

Figure 23:   Depth to Saline Ground Water (Geraghty et al, 1973)

11	Vertical intrusion caused by excessive pumping
8	Improper disposal of oil field brines
6	Intrusion caused by faulty well casings
5	Surface infiltration
5	Layers of salt water in thick limestone formations
2	Vertical intrusion caused by dredging
2	Irrigation return flow

Newport (1977) also provided a state-by-state summary which gives representative samples of salt-water intrusion; Table 8 includes the summary (Newport, 1977).

The second study was published by the U.S. Water Resources Council in 1981. The report provided a summary of ground water problems in each of the Council's Water Resource Regions. According to the report, all of the eighteen regions in the conterminous United States have experienced at least one type of ground water salinity problem. Some regions reported ground water quality problems due only to natural saline or "mineralized" water, while other regions cited specific examples of salt-water intrusion. Figure 24 shows the locations of the eighteen water resource regions (U.S. Water Resources Council, 1981). Table 9 contains a synopsis of all saline ground water problems listed for the eighteen regions.

NATIONAL OVERVIEW

The severity of salt-water intrusion in the conterminous United States ranges from severe problems to little or no threat in localized regions. When viewed as a national issue, however, the problem is quite wide spread. Based on an extensive search of the literature and the national telephone survey, Figure 25 indicates the states which are currently experiencing problems with salt-water intrusion. A total of 39 of the 48 contiguous states have indicated reduced ground water quality as a result of some form of salt-water intrusion. This conclusion is based on information gathered in earlier research efforts (see first section of this chapter), analysis of telephone contacts, and a concentrated search of the published literature.

A majority of the inland states in the central region of the nation are experiencing saline ground water issues. Although the problems are due to a variety of sources, it is not surprising since these states typically have shallow saline ground water (see Figure 23). Those states in the western half of the nation not experiencing salt-water intrusion, however, are faced with surface water salinity problems (see Figure 22).

Table 8: State-by-State Summary of Representative Problems of Salt-water Intrusion (Newport, 1977)

Location (1)	Nature of problem (2)	Corrective measures taken (3)	Outlook (4)
ALABAMA Mobile-Gulf Coast	Lateral intrusion from Mobile River caused by intensive pumping	Pumping curtailed; deeper wells for fresh water	Status quo for shallow aquifer
Marango County-Coastal Plain	Upward flow of saline water within a fault	Well field moved to safer location	Unknown
ALASKA Yakutat area-Gulf of Alaska	Lateral intrusion from the ocean on a narrow sand spit when pumping from a 70-ft vertical well (Ghyben-Herzberg principle)	Installed shallow infiltration gallary to skim fresh water from the lens overlying sea water	Okay if demand does not exceed supply
Cook Inlet area-Anchorage	Potential of lateral intrusion from the ocean caused by intensive pumping	None	Present contamination at Fire Island; hazard to Anchorage well field; monitoring wells to be installed
ARIZONA	No known examples		

Table 8: (Continued)

Location (1)	Nature of problem (2)	Corrective measures taken (3)	Outlook (4)
ARKANSAS Various	Potential contamination from oil field brines leaking into fresh water aquifers	State requires casing or plugging of wells	Under control
Eastern	Lateral movement of saline water because of pumping	None	Unknown, depends upon all factors in hydrologic system; study proposed
Southern	Lateral salt water intrusion caused by updip migration resulting from pumping	None	Gradual local encroachment
CALIFORNIA Ventura County-Oxnard Plain	Lateral intrusion from ocean caused by intensive pumping	Experimental facilities in operation for control with a pumping trough by State Department of Water Resources and United Water Conservation District	Economic pressure will force solution; experimental work is continuing

Table 8: (Continued)

Location (1)	Nature of problem (2)	Corrective measures taken (3)	Outlook (4)
Santa Clara County	Lateral intrusion from San Francisco Bay caused by intensive pumping	Pumping curtailed; recharging aquifer artificially	Managed ground-water basin
Los Angeles County-West Coast Basin	Lateral intrusion from ocean caused by intensive pumping	Intrusion stopped with a fresh water pressure barrier; pumping rates stabilized	Continued operation of barrier by Los Angeles County Flood Control District and management of the ground-water basin
COLORADO Denver Arsenal	Surface infiltration and lateral movement of industrial wastes caused contamination of adjacent aquifers	Industrial wastes moved to deep disposal well; well injection correlates with increased earthquake activity	Controversy
CONNECTICUT New Haven and Bridgeport	Lateral intrusion from tidewater in harbors caused by intensive pumping	Pumping relocated landward; alternate supplies used	Further pumping curtailment and greater use of alternate supplies

Table 8: (Continued)

Location (1)	Nature of problem (2)	Corrective measures taken (3)	Outlook (4)
DELAWARE Coastline and Delaware River	Lateral intrusion from tidal water in Delaware River and Bay and from ocean caused by intensive pumping and dredging of impermeable soils	Pumping relocated landward	Continued intrusion; further pumping curtailment
FLORIDA Dade and Broward Counties-Miami	Infiltration of tidal water from canals constructed to drain inland areas and to lower water table	Canal construction controlled; installed canal salinity control structures to keep out sea water and to raise level of fresh water	Continued management of factors affecting water supply; continued surveillance and studies
Pinellas County-St. Petersburg	Lateral intrusion from ocean and Tampa Bay into thick limestone aquifers caused by intensive pumping	Pumping reduced	Investigation and management effort underway

Table 8: (Continued)

Location (1)	Nature of problem (2)	Corrective measures taken (3)	Outlook (4)
Cocoa Beach– Cape Canaveral	Upward movement of residual salt water within thick limestone aquifer caused by intensive pumping	Pumping curtailment will be necessary. Replenishment by injecting surface water is being considered	Pumping limited to available supply
Hendry County– Southwest Florida	Localized upward movement of residual salt water into thick limestone aquifer caused by intensive pumping and broken or corroded well casings	Stopped pumping to the area	No change
GEORGIA Savannah area	Potential lateral intrusion from ocean into limestone aquifer resulting from massive cone of depression caused by intensive pumping	None yet because intrusion has not reached large production wells	Major pumping curtailment; continuing cooperative investigations

Table 8: (Continued)

Location (1)	Nature of problem (2)	Corrective measures taken (3)	Outlook (4)
Brunswick	Wells encounter layers of residual salt water in thick limestone aquifers	None reported	Pumping from selected zones only; amount of pumping probably will have to be controlled; continued cooperative investigations
HAWAII Oahu	Potential lateral intrusion from ocean where pumping is too deep or or excessive	Leaky wells controlled; pumping limited to amount of recharge, and locations chosen carefully	Continued management by Board of Water Supply; City and County of Honolulu; permanent operation of monitoring wells; continued studies to maximize safe production
IDAHO	No known examples		
ILLINOIS	Brine disposal	State control	
	Suspected lateral or upward movement of saline water		

Table 8: (Continued)

Location (1)	Nature of problem (2)	Corrective measures taken (3)	Outlook (4)
INDIANA Various	Brine disposal	State control	Generally good
Mt. Vernon-West Franklin	Upward (?) flow of saline water through fault zones into fresh-water aquifers due to pumping	None	Bleak; abandonment of wells necessary
IOWA (?)	Potential updip migration of saline water within thick aquifers due to intensive pumping	None	Localized problems only(?)
KANSAS Various	Brine disposal	State Board of Health controls	Controlled
Various	Potential infiltration of saline streamflow	None	Trouble; possible control of saline water sources
Various	Salt mine waste disposal	Disposal within mined out areas	Controlled

Table 8: (Continued)

Location (1)	Nature of problem (2)	Corrective measures taken (3)	Outlook (4)
KENTUCKY	Discharge of oil field brines into streams and subsequent infiltration	State controls	Gradual improvement
LOUISIANA Baton Rouge	Lateral movement of residual saline water (or possibly industrial waste) into water supply well field due to intensive pumping	None yet	Reduced pumping--supplementary water supply; continuing intensive studies
Vermilion River area	Lateral intrusion of tidal water from Vermilion River during low-flow periods into producing aquifer	Reduced pumping	River salinity control structure proposed
MAINE	No known examples		
MARYLAND Baltimore and Sparrow Point	Lateral and vertical intrusion from tidal estuary of Patapsco River into producing	Little; reduction in pumping	Continuing intrusion; additional abandonment of wells; development

Table 8: (Continued)

Location (1)	Nature of problem (2)	Corrective measures taken (3)	Outlook (4)
	aquifers; problem was aggravated by harbor dredging, which improved exposure of permeable materials, and by leaky and broken well casings, which conducted saline water to deeper foundations		of alternate supplies
MASSACHUSETTS Provincetown, Scituate and Somerset	Minor lateral intrusion from ocean and salt water marshes in shallow aquifers because of heavy pumping	None	Continuing local problem
MICHIGAN Various	Upward intrusion of saline water from deep bedrock into producing glacial aquifers due to pumping; sometimes	Reduced pumping and alternate water supply	Continued problem requiring adjusted pumping pattern or alternate water supply

Table 8: (Continued)

Location (1)	Nature of problem (2)	Corrective measures taken (3)	Outlook (4)
	aggravated by heavy pumping and/or leaky or broken well casings		
MINNESOTA Northwest	Upward intrusion of saline water from deep bedrock into producing glacial aquifers caused by pumping	Reduced pumping	Use of alternate supplies
MISSISSIPPI Pascagoula area	Pascagoula Formation is subject to intrusion of salt water moving updip from the Gulf of Mexico	None	No problems yet; cooperative monitoring program underway
MISSOURI Various	Potential upward insion into producing aquifers from deep, saline aquifers, if pumping draft becomes too heavy		Good, because problem recognized

Table 8: (Continued)

Location (1)	Nature of problem (2)	Corrective measures taken (3)	Outlook (4)
MONTANA Various	Brine disposal and leaky wells in saline formations	State laws	Trouble, compliance checks inadequate
NEBRASKA Northeast and East	Potential upward intrusion into producing aquifers from deep, saline aquifers	None	Not serious
NEVADA	No known examples; however, an inherent potential for lateral movement of saline groundwater		
NEW HAMPSHIRE Portsmouth	Minor lateral intrusion from tidal water in Piscataqua River	Unknown	Continuing problem
Various	Possible contamination by highway salt; this is a potential problem in		Alternate deicing methods when problem becomes serious; studies underway

Table 8: (Continued)

Location (1)	Nature of problem (2)	Corrective measures taken (3)	Outlook (4)
	many states, but apparently was not considered a salt water intrusion problem by most respondents		
NEW JERSEY Newark-Passaic River, Sayreville-Raritan River, Camden-Delaware River	Lateral intrusion from tidal estuaries into producing aquifers, aggravated by intensive pumping, harbor and canal dredging, and the disposal of industrial and municipal wastes	Pumping relocated; use of alternate supplies	Serious, until control measures established; studies are continuing
Atlantic City-Cape May	Lateral intrusion from ocean and Raritan and Delaware Bays due to pumping	Pumping moved landward	Continued intrusion
NEW MEXICO Various	Upward intrusion into producing aquifers from deep, saline	Principally re-location of pumping wells	Grim

Table 8: (Continued)

Location (1)	Nature of problem (2)	Corrective measures taken (3)	Outlook (4)
	bedrock formations because of heavy pumping		
NEW YORK Long Island	Lateral intrusion from ocean into producing aquifers caused by heavy pumping and reduced natural recharge	Artificial recharge of storm runoff; reduced pumping; use of alternate supplies; experiments with reclaimed water injection by USGS and Nassau County underway	Continuing intrusion; additional control measures and artificial recharge; intensive studies are continuing
NORTH CAROLINA Wilmington New Bern	Lateral intrusion from tidal estuaries into shallow producing aquifers due to heavy pumping	Use of alternate supplies	Unknown

Table 8: (Continued)

Location (1)	Nature of problem (2)	Corrective measures taken (3)	Outlook (4)
NORTH DAKOTA Red River Valley	Upward intrusion into producing aquifers from deep, saline bedrock formations due to pumping	None	Not good; studies underway
OHIO Muskingum River Basin	Industrial waste from chemical plants	Self-regulation by industry	Good
Various	Oil brine disposal	Regulation by state	Good
OKLAHOMA Various	Potential infiltration of oil field brines	State-controlled standards for deep disposal wells	Problems probably not increasing
OREGON	No known examples		
PENNSYLVANIA Philadelphia	Lateral intrusion of shallow aquifer by tidal water from Delaware River and	Using alternate supplies	Continued intrusion until control measures instituted

Table 8: (Continued)

Location (1)	Nature of problem (2)	Corrective measures taken (3)	Outlook (4)
	by infiltration of industrial and municipal wastes, aggravated by heavy pumping and harbor dredging		
RHODE ISLAND Providence Warren	Lateral intrusion of glacial outwash aquifers from tidal estuaries and ocean caused by pumping	Stopped pumping	Develop inland ground water supplies; artificial recharge in coastal areas possible
SOUTH CAROLINA Paris Island	Heavy pumping in limestone aquifers causes upward and downward intrusion from layered saline aquifers	Pumping reduced	Limited pumping only
Beaufort area	Lateral intrusion from ocean into shallow, producing aquifer due to pumping		

Table 8: (Continued)

Location (1)	Nature of problem (2)	Corrective measures taken (3)	Outlook (4)
SOUTH DAKOTA Black Hills	Potential updip intrusion of saline water into producing aquifer because of increased pumping	None yet	Pumping curtailment may be necessary
Various	Localized upward intrusion into producing aquifers from deep, saline formations due to pumping	None	
TENNESSEE	No known examples		
TEXAS Galveston-Texas City	Upward and/or downward intrusion of residual saline water into producing aquifers because of heavy pumping	Pumping moved inland; surface supplies developed; desalting being considered	Continuing problems

Table 8: (Continued)

Location (1)	Nature of problem (2)	Corrective measures taken (3)	Outlook (4)
UTAH Great Salt Lake area	Potential lateral intrusion from lake into producing aquifers because of heavy pumping	None yet	May be serious
Western	Potential upward and/ or lateral intrusion of saline waters into producing aquifers; also, an increase in salinity of ground water from irrigation return flow	None	Conditions getting worse
VERMONT	No known examples		
VIRGINIA Newport News Cape Charles	Contaminated wells; could be intrusion of seawater or residual saline water	Unknown	Not known
WASHINGTON Tacoma area	Lateral intrusion from ocean due to pumping	Wells moved inland	Continued intrusion

Table 8: (Continued)

Location (1)	Nature of problem (2)	Corrective measures taken (3)	Outlook (4)
Grant County	Lateral intrusion into producing aqui-from vicinity of saline lakes because of pumping	None	
WEST VIRGINIA	No known examples; however, an inherent potential for move-ment of residual saline water		
WISCONSIN Various	Lateral intrusion of saline water because of heavy pumping	None	Not generally serious
Various	Cannery waste disposal	State control	Serious, if control not effective
WYOMING Various	Mixing of fresh and saline water by intrusion through old oil wells	None	Localized problems

Table 8: (Continued)

Location (1)	Nature of problem (2)	Corrective measures taken (3)	Outlook (4)
Various	Salinity increasing in groundwater because of irrigation return flow	None	Minor problems

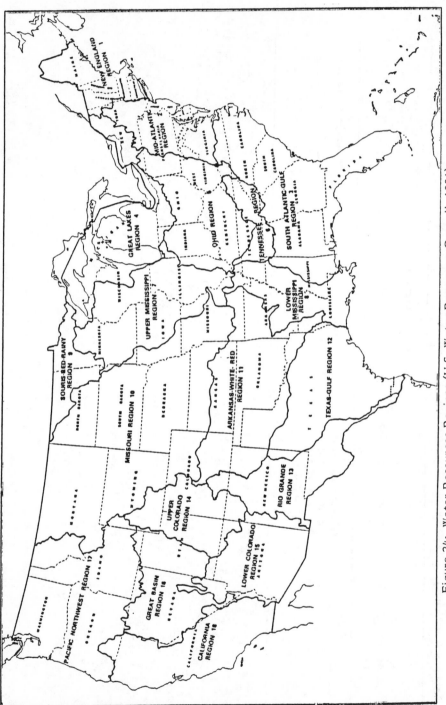

Figure 24: Water Resources Regions (U.S. Water Resources Council, 1981)

Table 9: Saline Ground Water Problems Identified by Water Resources
Region (U. S. Water Resources Council, 1981).

Region	Supply Problems	Natural Quality Problems	Man-made Quality Problems
New England (1)	Some saltwater intrusion along coastal areas.	High dissolved solids and sulfate in Triassic sedimentary rocks of central Massachusetts and Connecticut	According to a 1980 survey conducted for the New England Congressional Caucus, contamination of ground water is the most serious environmental problem facing the region. Salinity problems include: 1. Road salt contamination. 2. Localized saltwater intrusion.
Mid-Atlantic (2)	Overpumping is causing serious saltwater intrusion in places like Long Island and New York City. It appears that similar problems are beginning to occur in South Jersey and Norfolk, Virginia. Expansion and deepening of regional cone of depression in piezometric surface in coastal plains.	Naturally occurring saltwater in Coastal Plain aquifers of New Jersey and Delaware. High sulfates and total dissolved solids in Triassic rocks of southeastern Pennsylvania.	Localized saltwater intrusion.
South Atlantic Gulf (3)	Extensive cones of depression and lowered artesian heads over hundreds of square miles	Naturally occurring saline water.	Contamination from oil and natural gas field brines. Saltwater intrusion.

Table 9: (Continued)

Region	Supply Problems	Natural Quality Problems	Man-made Quality Problems
	due to large-scale concentrated ground water pumping and dewatering operations. Saltwater intrusion.		
Great Lakes (4)	Localized overpumping, principally in urban areas like Chicago and Milwaukee, has resulted in critically lowered ground water levels or induced migration of heavily mineralized water into freshwater aquifers.	Inferior chemical quality of ground water in many areas, made worse by man's activities.	Road salt contamination. Induced intrusion of heavily mineralized water into fresh water aquifers.
Ohio River (5)		In a few areas naturally inferior chemical quality of ground water, which is made worse by man's activities.	Contamination from oil and gas wells: - improper well construction, abandoned oil and gas production wells - brine disposal.
Tennessee (6)		Relatively shallow depth of mineralized ground water in parts of region.	Existence of open and abandoned test holes for zinc and oil exploration, which may permit brackish water to rise and contaminate freshwater zones.

Table 8: (Continued)

Region	Supply Problems	Natural Quality Problems	Man-made Quality Problems
Upper Mississippi (7)		In Northeastern Missouri, Western Illinois, and Southern Iowa, ground water yields are low and highly mineralized.	
Lower Mississippi (8)			Oil field brine disposal is a problem in some oil and gas producing areas. Saltwater intrusion is a potential problem in local coastal areas.
Souris-Red Rainy (9)		Excessive natural concentrations of total dissolved solids, sulfates, and chloride are common in the western portion of the region.	Contamination from mining activities associated with the production of taconite, coal, and oil or gas.
Missouri (10)		Large areas of region, particularly in North and South Dakota, have ground water of inferior chemical quality.	
Arkansas White-Red (11)		Chloride along Salt Fork in North Central Oklahoma.	Degradation of water quality in stream-valley alluvium caused by reuse of water for irrigation along Arkansas

Table 9: (Continued)

Region	Supply Problems	Natural Quality Problems	Man-made Quality Problems
		Total dissolved solids, chloride and sulfate in Northern New Mexico. High total dissolved solids in local areas of the Upper Red River.	River Valley from Pueblo, Colorado, to the Kansas border.
Texas-Gulf (12)	Water level declines and increased pumping costs due to lowering of artesian pressure. Areas of most significant decline include the Houston and Dallas-Fort Worth regions. Other problems related to overdrafting and lowering of artesian heads include: depletion of streamflow; saltwater intrusion; land surface subsidence.	Relatively high fluoride concentrations and locally high nitrate concentrations in minor aquifers.	Localized saline water intrusion. Brine disposal associated with oil and gas exploration and production.
Rio Grande (13)	Saline water intrusion.	Ground water withdrawn is becoming more saline as the result of saline water intrusion and the recycling of water used for irrigation.	

Table 9: (Continued)

Region	Supply Problems	Natural Quality Problems	Man-made Quality Problems
Upper Colorado (14)		Water logging and the resulting high mineralization of shallow ground water in the San Luis Valley.	
		Ground water is often high in total dissolved solids.	Increased exploration for oil and gas could result in brine pollution.
Lower Colorado (15)		Contamination from agricultural runoff.	
Great Basin (16)		Moderately mineralized ground water is found in most parts of basin.	Ground water pollution from irrigation return flows.
California (18)	Uncorrected saltwater intrusion in the Santa Cruz-Pajaro, Salinas, and Ventura Central Basins. Hydraulic barriers and other management practices have controlled sea water intrusion in other previously affected major basins.	Saline connate waters occur in parts of the Sacramento and San Joaquin Valleys, and in some Lahontan and Colorado Desert ground water basins. Isolated cases of locally high concentrations of various natural constituents also exist throughout California.	Degradation of ground water from agriculture runoff near natural recharge points. Uncorrected saltwater intrusion in the Santa Cruz-Pajaro, Salinas, and Ventura Central Basins.

Figure 25: States Which Are Currently Experiencing Problems with Salt-water Intrusion

The following section provides details on each state's salt-water intrusion and lists many sources of documentation.

STATE-BY-STATE SUMMARY

The following information indicates the current extent of knowledge of salt-water intrusion problems in the contiguous United States. Figure 26 contains the legend used for each map.

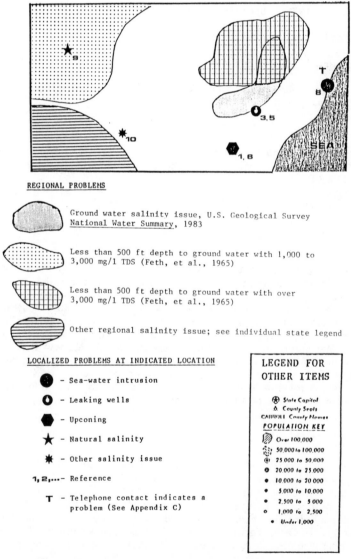

LEGEND FOR SALT-WATER INTRUSION

REGIONAL PROBLEMS

Ground water salinity issue, U.S. Geological Survey *National Water Summary*, 1983

Less than 500 ft depth to ground water with 1,000 to 3,000 mg/1 TDS (Feth, et al., 1965)

Less than 500 ft depth to ground water with over 3,000 mg/1 TDS (Feth, et al., 1965)

Other regional salinity issue; see individual state legend

LOCALIZED PROBLEMS AT INDICATED LOCATION

● - Sea-water intrusion

◑ - Leaking wells

⬣ - Upconing

★ - Natural salinity

✳ - Other salinity issue

1, 2,... - Reference

T - Telephone contact indicates a problem (See Appendix C)

LEGEND FOR OTHER ITEMS

⊕ *State Capitol*
△ *County Seats*
CANNON *County Homes*
POPULATION KEY

▨ *Over 100,000*
⠿ *50,000 to 100,000*
⊕ *25,000 to 50,000*
◐ *20,000 to 25,000*
● *10,000 to 20,000*
• *5,000 to 10,000*
• *2,500 to 5,000*
○ *1,000 to 2,500*
· *Under 1,000*

Figure 26: Legend for Salt-water Intrusion

*[a] 1.   Anderson, 1978 (AB-23)[b].   Upconing.

2.   Brown, 1981.   Brine disposal.

*   3.   Denver, 1983.   Leaking wells.

4.   Jones, 1975.   Lateral intrusion.

*   5.   Long, 1980.   Leaking wells.

*   6.   Morton, 1983 (AB-191).   Upconing.

7.   Nelson, 1975.   Upconing.

*   8.   O'Neal, 1977 (AB-55).   Sea-water intrusion.

*   9.   Parker, 1981.   Natural salinity.

*   10.   U.S. Environmental Protection Agency, 1979 (AB-140).
Brine disposal.

---

a.   *   Location indicated on map.
b.   (AB-##) Report is summarized in the Annotated Bibliography
(Appendix A).

Figure 26:   (Continued)

Alabama

Alabama (Figure 27) contains a large area with shallow, moderately saline ground water in the northern half of the state and two smaller pockets of moderately saline ground water in the coastal plain. Two areas of highly saline ground water exist at less than 500 feet in depth, one in the southwestern portion of the state and one along the northern state border. Several reports have addressed salt-water intrusion in Alabama; two specifically address upconing (in Marengo and Mobile Counties); one describes how fresh ground water "down-dips" into saline ground water (a natural phenomenon occurring in Montgomery County), and one describes sea-water intrusion (from Fort Morgan to Gulf Shores). A telephone survey also indicated that salt-water intrusion has been and continues to be a problem in Alabama.

Arizona

Arizona as a state is experiencing increasing salinity. Although the problem is primarily associated with surface water (pollution from agricultural runoff and the increasing salinity of the Colorado River), several ground water basins are also increasing in salinity. Figure 28 indicates locations where ground water salinity has been recognized as an issue (around the cities of Kingman, Holbrook, Camp Verde, Wilicox, and Sells, and several pockets around Phoenix). Furthermore, several areas in both the northeast and southwest

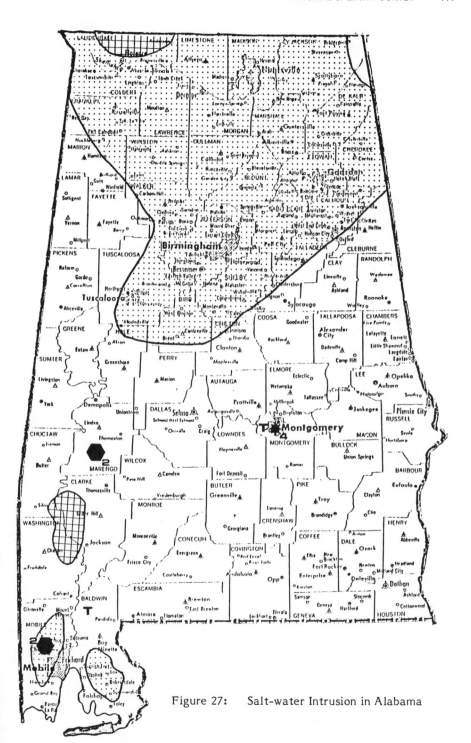

Figure 27:   Salt-water Intrusion in Alabama

1. Miller, et al., 1977. Salt-water intrusion.

* 2. Task Committee on Saltwater Intrusion, 1969, and Newport, 1977.
     Lateral intrusion and upconing.

3. Todd, 1983. Saline water overlying fresh water.

* 4. Todd, 1983. Fresh water 'downdips' into saline water.

5. Walter, Kidd, and Lamb, 1979 (AB-43).   Sea-water intrusion,
   leaking aquifers, lateral intrusion.

* T. Problem identified via telephone contact.

Figure 27:   (Continued)

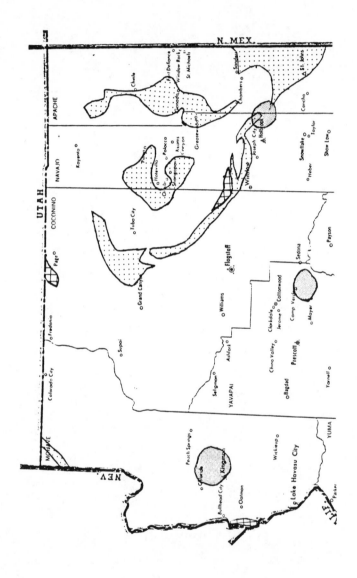

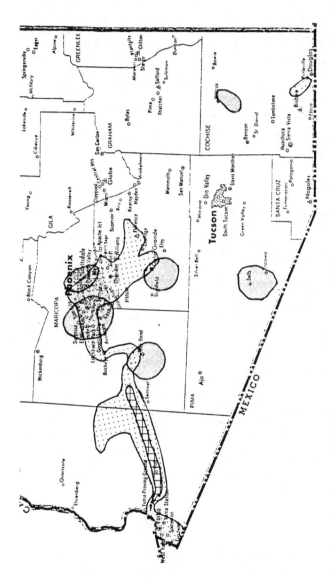

Figure 28:   Salt-water Intrusion in Arizona

1. Daniel, 1981. Map showing the location of saline water.

2. U.S. Water Resources Council, 1981.   Contamination from
   agricultural runoff.

regions of the state have shallow saline ground water, increasing the potential for future salt-water intrusion.

Arkansas

There are three zones of shallow saline ground water in Arkansas (Figure 29); one surrounding Fort Smith, one in the very southeastern corner of the state, and a narrow zone which reaches from the southwestern corner of the state through Little Rock and up into Monroe and Jackson Counties. Although there have not been many reported problems associated with these zones of shallow saline ground water, excessive ground water pumpage in these areas could result in salt-water intrusion. However, salt-water intrusion is a recognized issue in Arkansas: the disposal of brine and brackish waters which are produced with oil and gas field activities have deteriorated the quality of several ground water basins in Arkansas. Increasing ground water salinity has been documented in areas such as Independence, White, Woodruff, Chicot, Ashley, and Union Counties.

California

Sea-water intrusion has been a documented problem in the major coastal cities of California for many years (see Figure 30). Steadily increasing water demands associated with the increasing population in cities such as Los Angeles and San Francisco have accelerated the rate of ground water withdrawal, inducing sea-water intrusion. Smaller cities also, such as Santa Cruz and Salinas, have had sea-water intrusion reported. However, California has aggressively addressed the problem (especially in areas such as Orange County), and several reports have described various techniques for controlling sea-water intrusion. Techniques ranging from artificial injection, for creating a hydraulic barrier, to water desalination have been applied in California's coastal areas with varying degrees of success.

In addition to sea-water intrusion, California contains several inland areas which may soon be affected by salt-water intrusion. A large portion of the San Joaquin Valley and several other areas in the southern half of the state are underlain by shallow saline ground water. Due to large irrigation water uses, the shallow saline water has begun to intrude into the fresh water aquifers. The problems have been addressed using techniques ranging from planting salt tolerant crops to rationing. Research in California has shown, however, that the most successful method for controlling salt-water intrusion is in an integrated ground water management program which can address multiple problems, rather than single ground water issues.

Colorado

Salt-water intrusion in Colorado can be considered a minor problem. Although a fairly significant portion of the state is underlain by shallow saline ground water (see Figure 31), it appears that few problems have arisen. However, contamination from agricultural runoff is a problem in the Arkansas River Valley and the Colorado River Valley. Furthermore, since the southeastern part of the state is underlain by shallow saline ground water and

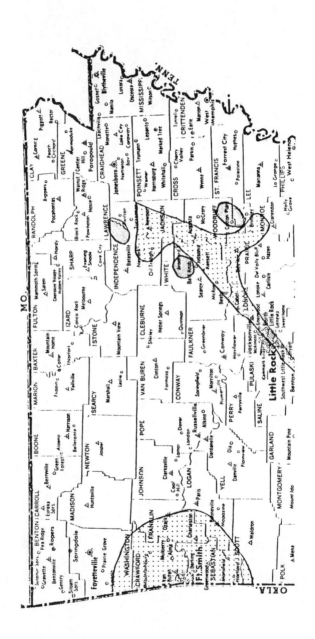

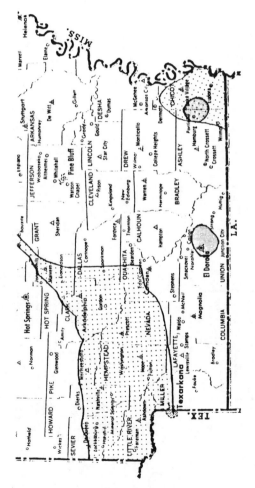

Figure 29:   Salt-water Intrusion in Arkansas

1. Newport, 1977.   Lateral intrusion in the eastern part.   Oil field brines.

2. U.S. Water Resources Council, 1981.   Oil field brine disposal.

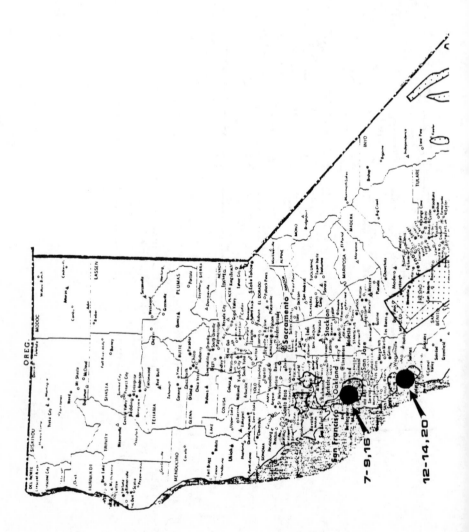

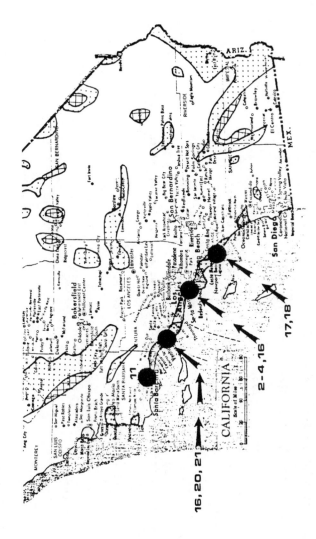

Figure 30:   Salt-water Intrusion in California

1. Anonymous, 1985 (AB-6). Sea-water intrusion.

* 2. Bookman, 1968 (AB-77). Sea-water intrusion, Los Angeles County.

* 3. Bruington, 1968 (AB-65). Sea-water intrusion, Los Angeles County.

* 4. Bruington, 1969 (AB-66). Sea-water intrusion, Los Angeles County.

5. Department of Water Resources, 1975.

6. Bruington, 1972 (AB-7). Sea-water intrusion.

7. Fractor, 1983 (AB-78). Property rights.

* 8. Jackson and Paterson, 1977 (AB-81). Sea-water intrusion, San Francisco.

* 9. Jeffers, et al., 1982 (AB-68). Sea-water intrusion, Santa Clara.

*10. Jenks and Harrison, 1977 (AB-69). Sea-water intrusion, Santa Clara.

11. Johnson, 1978 (AB-25). Sea-water intrusion modeling.

*12. Martin, 1982 (AB-28). Sea-water intrusion, Santa Barbara.

Figure 30:   (Continued)

*13. McDonald and Grefe, Inc., 1978 (AB-83). Water management in Monterey and Santa Cruz.

*14. Muir, 1974 (AB-73). Sea-water intrusion, Santa Cruz and Monterey.

*15. Muir, 1980 (AB-32). Sea-water intrusion, Santa Cruz.

16. Muir, 1982. Natural salinity.

*17. Newport, 1977. Sea-water intrusion, Ventura, Santa Clara, and Los Angeles.

*18. Orange County Water District, 1984. Net basin salt balance.

*19. Owen, 1968 (AB-74). Water management in Orange County.

20. Todd, 1983. Sea-water intrusion.

*21. U.S. Water Resource Council, 1981. Sea-water intrusion, Santa Cruz-Pajaro, Salinas, and Ventura. Natural salinity and degradation of ground water from agricultural runoff.

*22. Wedding, Fong, and Crafts, 1979 (AB-76). Sea-water intrusion, Oxhard Plain.

T. Problem identified via telephone contact.

Figure 30:   (Continued)

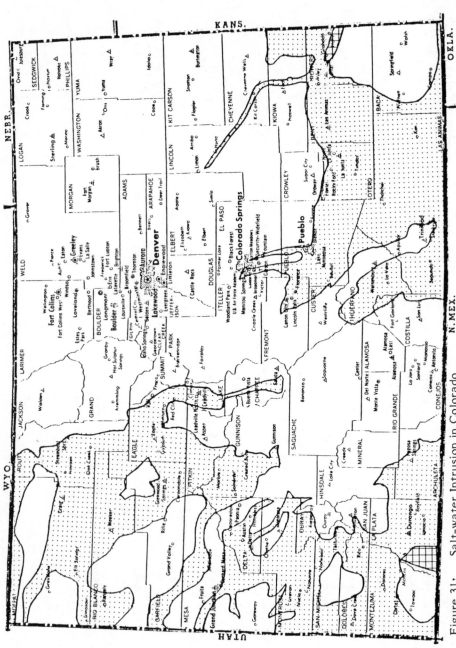

Figure 31:   Salt-water Intrusion in Colorado

1. U.S. Water Resources Council, 1981. Contamination from agricultural runoff, Arkansas River Valley.

Figure 31:   (Continued)

there is significant agricultural activity in the area, salt-water intrusion may be induced in the future.

Connecticut

Various reports have documented the existence of sea-water intrusion in Connecticut. Major coastal cities such as Bridgeport and New Haven have experienced lateral intrusion of sea-water which is caused by overpumping of fresh water aquifers. In fact, one source at the Institute of Water Resources, University of Connecticut, indicated that "...in every coastal town, private supplies show salt-water problems." However, sea-water intrusion is the only type of salt-water intrusion problem currently documented in Connecticut (see Figure 32).

Delaware

Salt-water intrusion is recognized as a major ground water quality issue in Delaware (Figure 33). New Castle, near the north end of Delaware Bay, has experienced sea-water intrusion primarily as a result of high industrial water demands on the local ground water. Reports have also indicated that all along Delaware's coast, sea-water intrusion has occurred. A telephone contact indicated that Sussex County's coastal cities consider salt-water intrusion a major problem, and records show that 1 to 2 public wells and 20 to 30 domestic wells are abandoned each year.

Florida

The entire Atlantic and Gulf Coast of Florida has experienced sea-water intrusion (Figure 34). The water demands of major cities such as Jacksonville, Tampa, and Miami have created large cones of depression in local aquifers, allowing sea-water to move inland. Cities with much smaller populations (e.g. Tarpon Springs and Cocoa Beach) are also creating conditions which induce sea-water to intrude into fresh water aquifers. Various other salt-water intrusion problems have also been documented in Florida. For example, leaking wells have occurred in Hendry, Nassau, Lee, and St. Johns Counties. Upconing has been documented in Hendry, Brevard, Martin, and Nassau Counties. Several reports have discussed modeling salt-water intrusion in Florida, control techniques used, and the legal aspects of salt-water intrusion.

Georgia

Salt-water intrusion is a major problem in two areas in Georgia (Figure 35): in Chatham County where Savannah's increasing population is the driving force for sea-water intrusion and around Brunswick in Glynn County where the high demand of three industries has created a large cone of depression inducing intrusion. Originally, the solution was suggested to involve dispersion of the well fields of these three industries. Economic reasons, however, prompted the industries to use extraction wells located just inland and to pump out sea-water in an attempt to prevent the sea-water from reaching the fresh water.

Although partially successful, the industries now practice various water

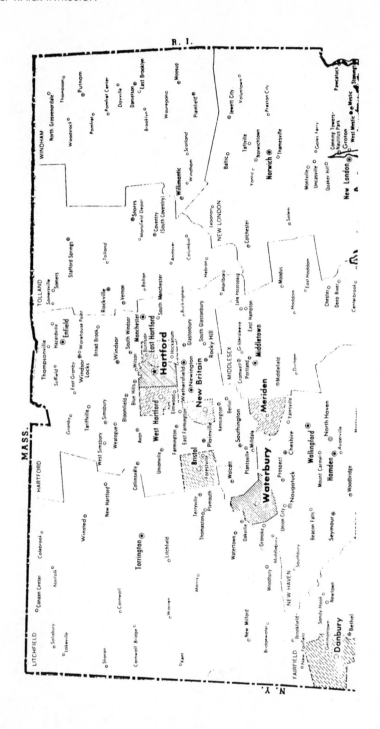

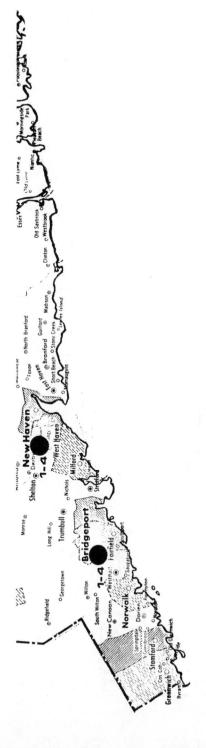

Figure 32: Salt-water Intrusion in Connecticut

* 1. Miller, DeLuca, and Tessier, 1974. Lateral intrusion of salt water due to overpumping.

* 2. Newport, 1977. Lateral intrusion of tidal waters due to overpumping.

* 3. Task Committee on Saltwater Intrusion, 1969. Lateral intrusion of tidal waters due to overpumping.

* 4. Todd, 1983. Lateral intrusion of salt water due to overpumping.

T. Problem identified via telephone contact.

Figure 33:   Salt-water Intrusion in Delaware

1. Groot, 1983. Sea-water intrusion in the coastal plain.

* 2. Miller, 1971. Sea-water intrusion due to overpumping.

3. Miller, DeLuca, and Tessier, 1974.   Sea-water intrusion along the coastline and Delaware River.

4. Newport, 1977.   Lateral   intrusion   of   sea-water   due   to overpumping.

* 5. Stegner, 1972(AB-92). Salt-water intrusion due to overpumping.

6. Task   Committee   on   Saltwater   Intrusion,   1969.   Sea-water intrusion along the coastline and Delaware River.

7. U.S. Water Resources Council, 1981.   Natural salt-water in the coastal plain.

* 8. Woodruff, 1969. Salt-water intrusion due to overpumping.

* T. Problem identified via telephone contact.

Figure 33:   (Continued)

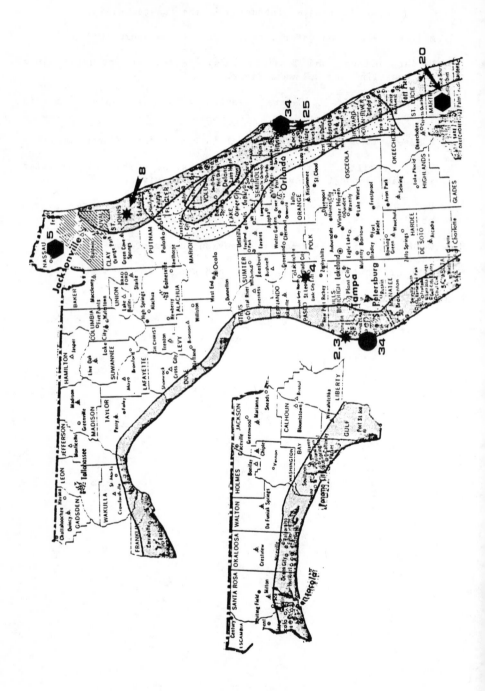

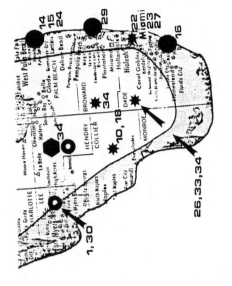

Figure 34: Salt-water Intrusion in Florida

* 1. Boggess, Missimer, and O'Donnell, 1977 (AB-48). Leaking wells.

* 2. Cherry, 1980 (AB-67). Pumping patterns.

* 3. Christensen, 1978 (AB-96). Leaking aquifer.

4. Christensen and Evans, 1974 (AB-86). Modeling salt-water intrusion.

* 5. Fairchild and Bentley, 1977 (AB-21). Upconing.

6. Florida House of Representatives, 1964 (AB-10). Salt-water intrusion.

7. Florida House of Representatives, 1964 (AB-11). Salt-water intrusion.

* 8. Franks and Phelps, 1979 (AB-50). Leaking aquifer.

  9. Grafton, 1968 (AB-22). Salt-water intrusion.

*10. Haire, 1979 (AB-23). Salt-water intrusion.

 11. Helwig, 1969 (AB-79). Legal aspects of salt-water intrusion.

 12. Hughes, 1979 (AB-24). Pressure trough for control.

 13. Klein, et al., 1975. Sea-water intrusion, tidal intrusion in canals, and leaking wells (South Florida).

*14. Land, 1977 (AB-27). Sea-water intrusion.

*15. Land, 1975 (AB-100). Modeling sea-water intrusion.

*16. Lee and Cheng, 1974 (AB-101). Sea-water intrusion.

 17. Maloney, 1955 (AB-82). Legal aspects of salt-water intrusion.

*18. McCoy, 1972 (AB-29). Salt-water intrusion.

 19. McCoy and Hardee, 1970 (AB-30). Sea-water intrusion (Southeastern Florida).

*20. Merritt, et al., 1983. Upconing.

 21. Miller, et al., 1977. Salt-water intrusion.

*22. Park, 1968 (AB-33). Salt-water intrusion.

*23. Pinder, 1975 (AB-106). Modeling sea-water intrusion.

Figure 34:   (Continued)

*24. Rodis and Land, 1976 (AB-34). Sea-water intrusion.

*25. Schiner, 1980 (AB-36). Salt-water intrusion.

*26. Science and Public Policy Program, 1983. Salt-water intrusion.

*27. Segol and Pinder, 1976 (AB-108). Modeling sea-water intrusion.

28. Sherwood and Klein, 1963. Sea-water intrusion and tidal intrusion in canals (Southeast Florida).

*29. Sherwood, McCoy, and Galliher, 1973. Sea-water intrusion.

*30. Sproul, Boggess, and Woodward, 1972. Leaking wells.

31. Stewart, et al., 1982. Mapping salt/fresh water interface.

32. Steinkampf, 1982. Upconing (Southwest coastal Florida).

*33. Swayze, 1979 (AB-41). Salt-water intrusion.

*34. Task Committee on Saltwater Intrusion, 1969, and Newport, 1977. Tidal intrusion in canals, sea-water intrusion, upconing, and leaking wells.

35. Todd, 1983. Saline water resources.

36. Ward, 1972 (AB-16). Salt-water intrusion (Everglades).

37. Wilson, 1982 (AB-45). Salt-water intrusion (West Central Florida).

T. Problem identified via telephone contact.

Figure 34: (Continued)

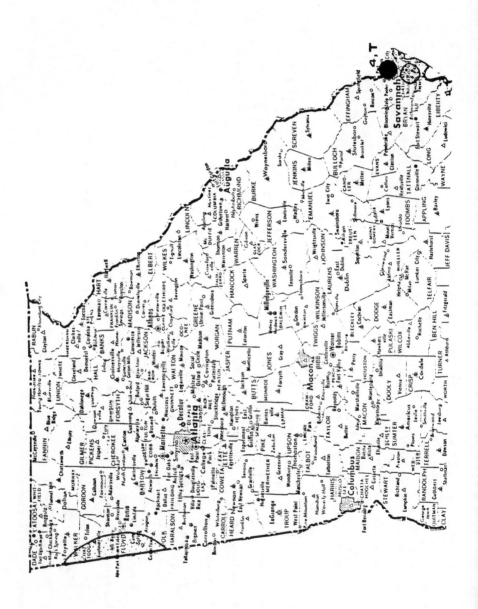

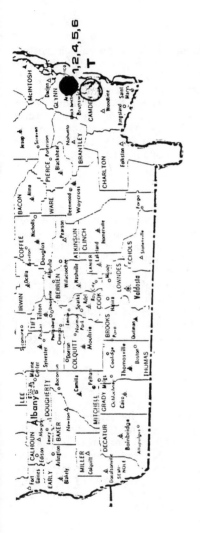

Figure 35:   Salt-water Intrusion in Georgia

* 1. Aral, Sturm, and Fulford, 1981 (AB-46).   Salt-water intrusion.

* 2. Gregg, 1971. Sea-water intrusion.

  3. Miller, et al., 1977. Natural salinity.

* 4. Task Committee on Saltwater Intrusion, 1969, and Newport, 1977.
       Sea-water intrusion.

* 5. Todd, 1983. Piezometric equilization.

* 6. Wait and Gregg, 1973. Sea-water intrusion.

* T. Problem identified via telephone contact.

conservation techniques which have resulted in slowing the intrusion. The only other potential problem area is in the northwestern part of the state where portions of Dade, Walker, Chattooga, Floyd, and Polk Counties overlie a region of shallow saline water.

Idaho

Idaho has not experienced major problems with salt-water intrusion. There have been a few localized incidents reported, primarily associated with the use of disposal and drain wells in the Snake River Plain. The reports indicate that ground water quality has deteriorated in areas around these disposal wells. There are, however, a few locations where shallow saline water exists (Figure 36), and these may be the sources of problems in the future. Jefferson, Bonneville, Oneida, Twin Falls, Owyhee, Canyon, and Payette Counties overlie pockets of moderately saline water, but Oneida and Bonneville Counties also have areas where the salinity exceeds 3000 mg/l TDS.

Illinois

The majority of Illinois is underlain by shallow moderately saline water (Figure 37), but only a few salt-water intrusion problems have been documented. Upconing and brine disposal have been mentioned as the most serious salt-water intrusion problems experienced in Illinois. However, due to the extent of shallow saline ground water reaching from a line which crosses nearly east to west through Peoria and covers almost all of the state south of the line, future salt-water intrusion problems are a distinct possibility. A salt-water intrusion study is just beginning in McHenry, Kane, Du Page and Cook Counties. Furthermore, a contact in the U.S. Geological Survey suspects that the saline Mt. Simon Aquifer may be leaking salt-water into shallower fresh water aquifers due to overpumping.

Indiana

Except for the extreme northwestern corner of the state, Indiana is completely underlain by shallow saline ground water ranging from 1000 to 3000 mg/l TDS (Figure 38). The Indiana Department of Natural Resources indicated, however, that most of the saline water is at least 500 feet deep, and ample rainfall prevents much overpumping which reduces the potential for salt-water intrusion. However, one report has documented the existence of a leaking aquifer in Knox County, and another report indicates that both upconing and brine disposal have caused salt-water intrusion in Posey County.

Iowa

Iowa is one of the eight states in the conterminous United States which is not currently experiencing salt-water intrusion. However, as Figure 39 indicates, the potential for serious problems in the future does exist. Well over one half of the state lies over shallow moderately saline water; however, four pockets of highly saline water also occur at depths less than 500 feet. These four pockets of saline water with more than 3000 mg/l TDS underlay all or parts of 24 of Iowa's 99 counties. Several reports have suggested the potential for

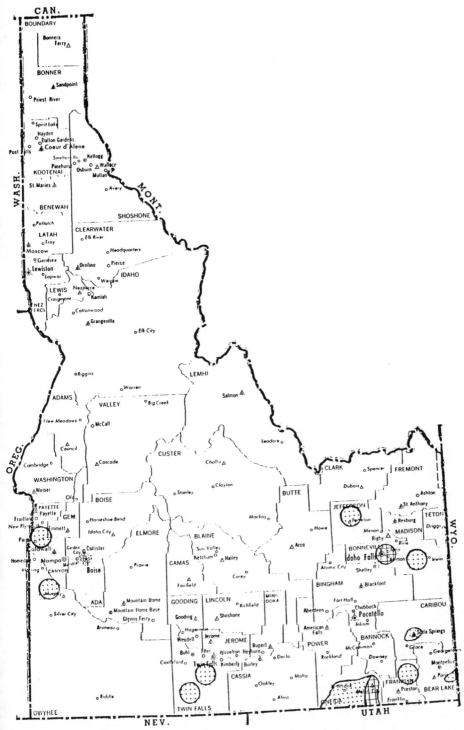

Figure 36:   Salt-water Intrusion in Idaho

1. Abegglen, 1970. Disposal wells.

2. Miller, 1980. Disposal wells.

3. Task Committee on Saltwater Intrusion, 1969, and Newport, 1977.

4. Todd, 1983. Saline water resources in southern areas.

5. U.S. Water Resources Council, 1981. Localized salt-water intrusion.

Figure 36:   (Continued)

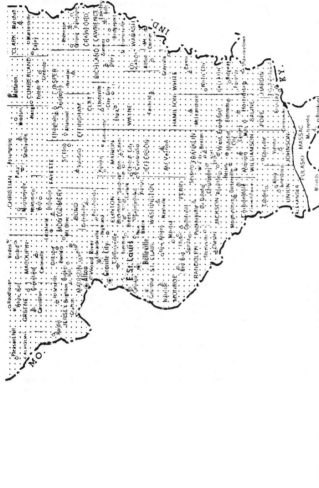

Figure 37:   Salt-water Intrusion in Illinois

1. Newport, 1977. Upconing and brine disposal.

2. Todd, 1983. Depth to saline aquifers.

3. U.S. Water Resources Council, 1981. Natural salinity.

South Bend · Fort Wayne · Hammond · Kokomo · Richmond · Muncie · Anderson · Indianapolis · Terre Haute · VERMILLION

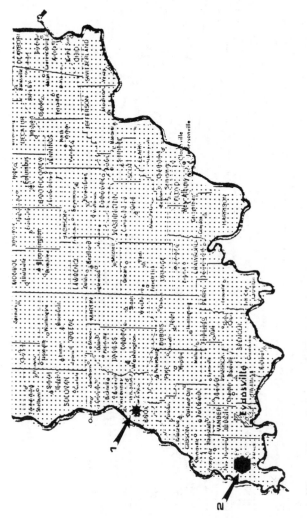

Figure 38: Salt-water Intrusion in Indiana

* 1. Shedlock, 1980 (AB-52). Leaking aquifer.

* 2. Task Committee on Saltwater Intrusion, 1969, and Newport, 1977. Brine disposal, upconing (Mt. Vernon).

T. Problem identified via telephone contact.

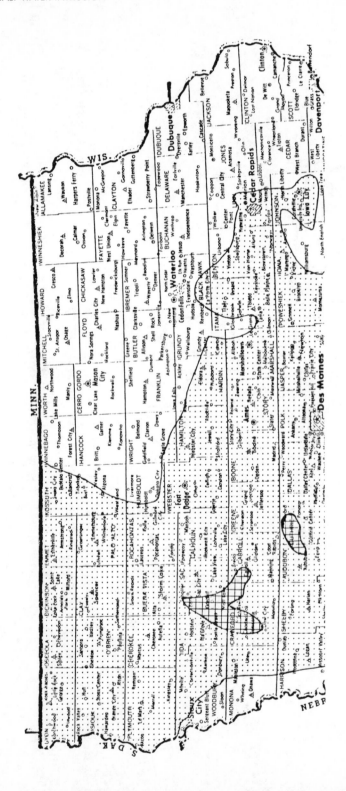

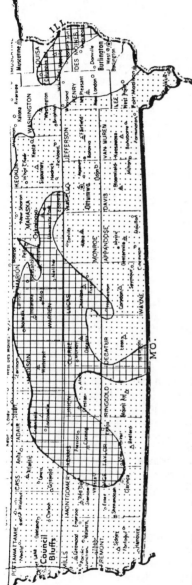

Figure 39: Salt-water Intrusion in Iowa

1. Burkart, 1984. Availability and quality of ground water (Northwestern).

2. Meyer and Barwell, 1983. Ground water resources.

3. Miller, 1980. Well disposal practices.

4. Task Committee on Saltwater Intrusion, 1969. Potential migration of saline water.

5. Todd, 1983. Saline ground water resources.

6. U.S. Water Resources Council, 1981. Natural salinity problem (Southern).

salt-water intrusion problems in Iowa, especially in southern areas.

Kansas

The existence of salt-water intrusion has been documented in Sedgwick, Reno, Harvey, and McPherson Counties, while the Kansas State Board of Agriculture mentioned Saline, Dickinson, Seward, and Meade Counties as locations. These counties make up the Solomon and Smokey Hill Rivers in the Wellington Formation where salinity problems are "evident". Furthermore, the oil and gas activity around Salina is suspected of increasing the salt-water intrusion problems in the area. The eastern two thirds of Kansas also overlies a large area of shallow moderately saline ground water and several pockets of highly saline water less than 500 feet deep (Figure 40).

Kentucky

Although almost the entire state overlays shallow saline water (Figure 41), the salt-water intrusion problems which exist in Kentucky are not generally associated with overpumping. Many problems are attributed to oil and gas field activities, including both oil and gas production and brine disposal. The Kentucky Geological Survey reports that natural saline ground water may become a problem, but little information is currently available.

Louisiana

Salt-water intrusion is a major problem in Louisiana. Various reports have described the intrusion which has occurred all along the coastal shores of Louisiana. The major cities of Lake Charles, Baton Rouge, and New Orleans, although located well inland, have been forced to greatly reduce the use of ground water because of the severity of salt-water intrusion. However, not only must Louisiana contend with sea-water intrusion along the Gulf Coast, there are other sources of saline water which must be considered. There are two large regions of salt domes in the southern portion of the state and one in the northern portion. Oil and gas field activities in these areas increase the potential for salt-water intrusion. These regions of salt domes are closely related to the shallow saline ground water covering approximately one half of the state (Figure 42). Finally, a zone of saline water in the Mississippi River alluvial aquifer has been mapped along the northeastern state border. These numerous sources of salt water indicate that unless steps are taken now to plan for control programs, intrusion will likely increase before it decreases in Louisiana.

Maine

The occurrence of salt-water intrusion has been documented on the coast of Maine (see Figure 43). South of Augusta, in Kennebec and Sagadahoc Counties, overpumping has induced the intrusion of the tidal estuary waters into the local aquifers. Maine's Division of Health Engineering--Drinking Water Program indicated that salt-water intrusion has been a problem for domestic wells and that the problem fluctuates with the time of year. Fortunately, other than localized salt-water intrusion problems along the coast, Maine should not

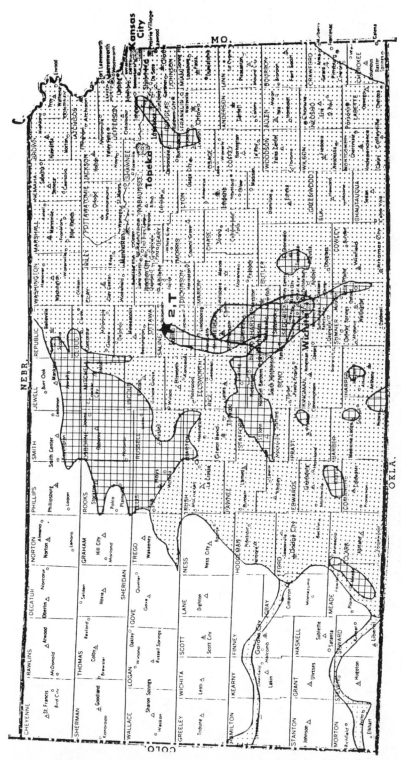

Figure 40:   Salt-water Intrusion in Kansas

1. Balsters and Anderson, 1979 (AB-62). Contamination from agricultural runoff and upconing.

* 2. McElwee, et al., 1984. Salinity control, Salina.

3. Newport, 1977. Natural salinity and brine and mine waste disposal.

4. Sloan, 1979 (AB-14). Ground water management.

5. Todd, 1983. Saline aquifer locations, well yield, and maximum Total Dissolved Solids.

* T. Problem identified via telephone contact.

Figure 40:   (Continued)

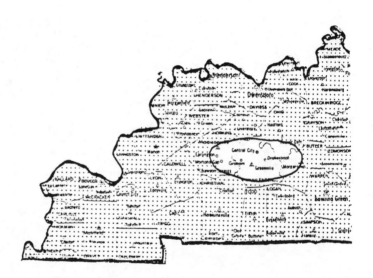

Figure 41:   Salt-water Intrusion in Kentucky

1. Task Committee on Saltwater Intrusion, 1969, and Newport, 1977. Oil field brines.

T. Problem identified via telephone contact.

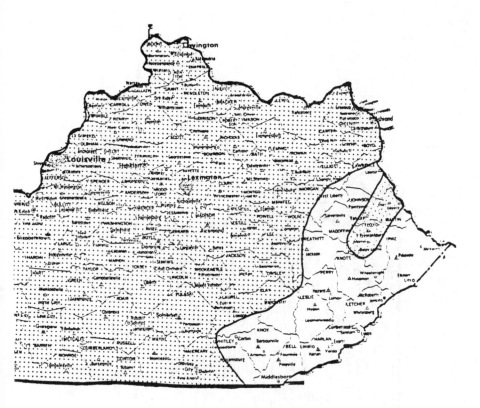

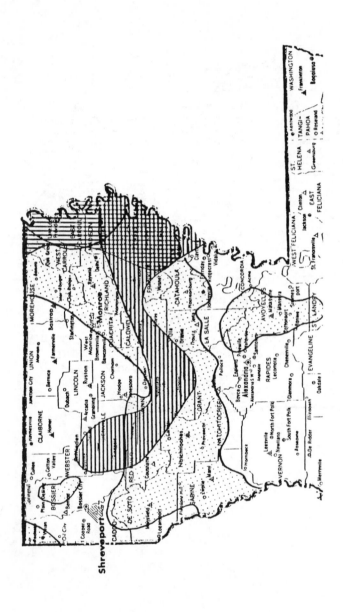

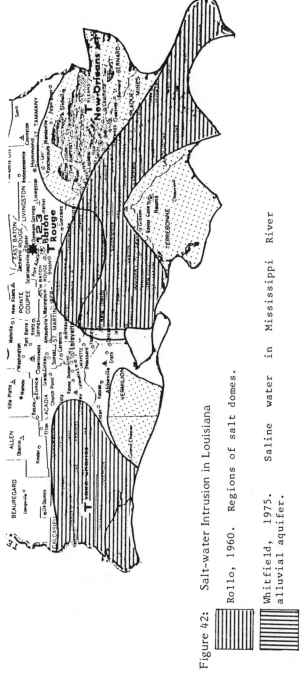

Figure 42:   Salt-water Intrusion in Louisiana

Rollo, 1960. Regions of salt domes.

Whitfield, 1975.   Saline water in Mississippi River alluvial aquifer.

* 1. Rollo, 1969. Upconing. Lateral intrusion.

* 2. Smith, 1969 (AB-38). Salt-water intrusion.

* 3. Task Committee on Saltwater Intrusion, 1969, and Newport, 1977. Lateral intrusion.

* T. Problem identified via telephone contact.

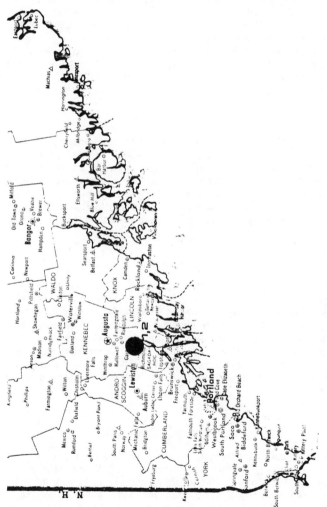

Figure 43:   Salt-water Intrusion in Maine

* 1. Miller, DeLuca, and Tessier, 1974. Salt-water intrusion from
      tidal river estuary due to overpumping.

* 2. Todd, 1983. Salt-water intrusion from tidal river estuary due
      to overpumping.

expect to face major problems with intrusion since there are no inland areas which overlie shallow saline ground water.

Maryland

Several areas along the Chesapeake Bay have experienced salt-water intrusion (Figure 44). St. Mary's, Anne Arundel, Harfork, Dorchester, and Somerset Counties have all experienced sea-water intrusion due to overpumping, leaky well casings, and dredging activities. Maryland's Division of Water Supply--Department of Health and Mental Hygiene indicated that salt-water intrusion had occurred at Kent Island but did not significantly affect drinking water. Control measures are being taken to reduce pumping rates in order to stop further intrusion. Other than sea-water intrusion, however, Maryland does not contain shallow saline ground water and should not expect future problems with salt-water intrusion away from its coastal shores.

Massachusetts

The three southernmost mainland counties in Massachusetts, Bristol, Plymouth, and Barnstable, as well as Essex and Dukes Counties, have had sea-water intrusion documented (Figure 45). Maine's Department of Environmental Quality Engineering indicated that Cape Cod in Barnstable County has partially controlled the sea-water intrusion problems by regulating pumping rates. However, Massachusetts does not have inland areas of shallow saline ground water and should not become affected by any other type of salt-water intrusion.

Michigan

It has been reported that oil field brines, upconing, and leaky wells have been responsible for salt-water intrusion in Michigan. Furthermore, a majority of the state overlies shallow saline ground water (Figure 46). A long narrow zone along the eastern edge of the state, beginning just north of Saginaw Bay and extending south to Maumee Bay, contains saline water with greater than 3000 mg/l TDS at depths less than 500 feet. If ground water resources are developed in response to water demands in Detroit, salt-water intrusion may occur. However, there have only been a few problems with salt-water intrusion documented in Michigan.

Minnesota

Salt-water intrusion, primarily upconing, is a recognized problem in Minnesota. A narrow zone along the western boundary of Minnesota, which widens to almost one half of the state at the southern boundary, as well as some isolated locations along the shores of Lake Superior have experienced salt-water intrusion. These areas of documented intrusion are highly related to the shallow salt-water aquifers in Minnesota (Figure 47). Intrusion problems can be expected to continue in the areas currently experiencing them as long as ground water use is high. Furthermore, the counties of Renville, Shelby, Nicollet, Brown, Blue Earth, and Fairibault, which are not currently experiencing problems with salt-water intrusion, may begin to develop intrusion situations unless ground water management programs are developed. These counties

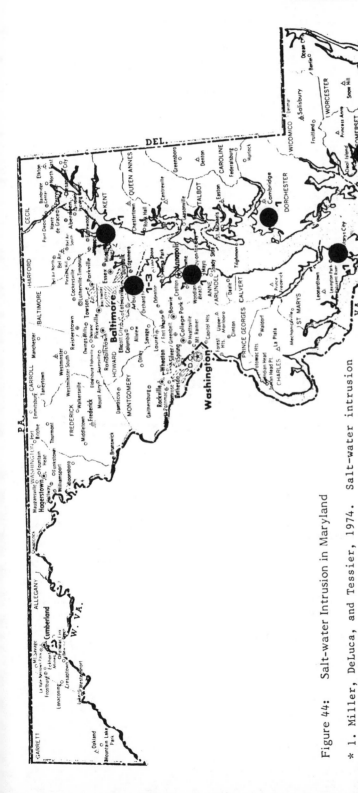

Figure 44: Salt-water Intrusion in Maryland

* 1. Miller, DeLuca, and Tessier, 1974. Salt-water intrusion due to overpumping.

* 2. Newport, 1977. Salt-water intrusion due to overpumping, leaky casings, and dredging.

* 3. Task Committee on Saltwater Intrusion, 1969. Salt-water intrusion due to overpumping, leaky casings, and dredging.

T. Problem identified via telephone contact.

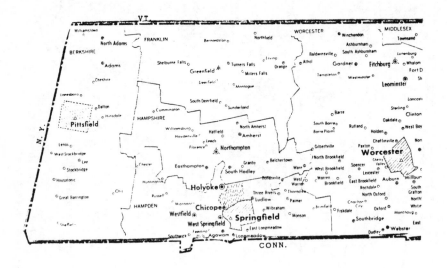

Figure 45:   Salt-water Intrusion in Massachusetts

* 1. Miller, DeLuca, and Tessier, 1974.   Lateral intrusion of sea water due to overpumping.

* 2. Newport, 1977.   Lateral intrusion of sea water due to overpumping.

* 3. Task Committee on Saltwater Intrusion, 1969.   Lateral intrusion of sea water due to overpumping.

* 4. Todd, 1983. Lateral intrusion of sea water due to overpumping.

  T. Problem identified via telephone contact.

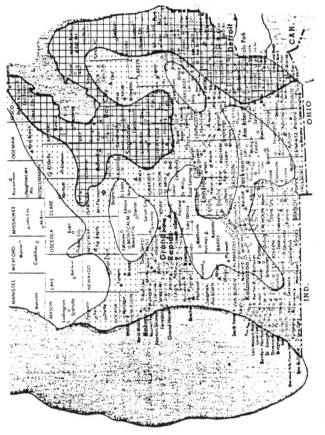

Figure 46:   Salt-water Intrusion in Michigan

1. Miller, 1980. Oil field brine contamination.

2. Task Committee on Saltwater Intrusion, 1969. Upconing and leaky wells.

T. Problem identified via telephone contact.

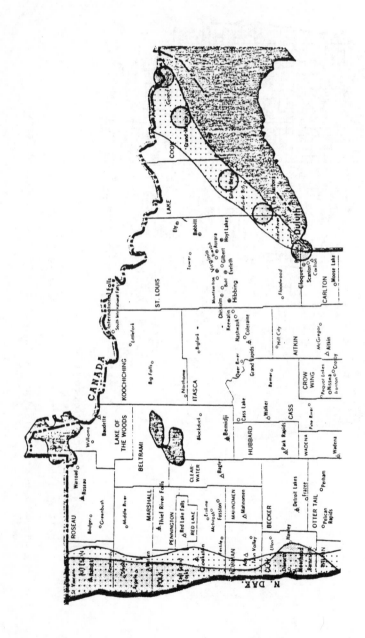

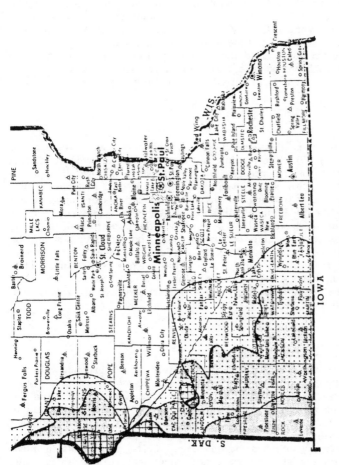

Figure 47:   Salt-water Intrusion in Minnesota

1. Task Committee on Saltwater Intrusion, 1969.   Upcoming (Northwest).

2. U.S. Water Resources Council, 1981. Natural salinity problem.

overlie the same zone of shallow moderately saline water which is currently creating problems in neighboring counties, thus indicating the potential for future problems.

Mississippi

Mississippi has experienced several cases of salt-water intrusion including sea-water intrusion, upconing, and oil and gas field activities. Two areas of sea-water intrusion have been documented, one in Hancock County and one in Jackson County. Two large zones and one smaller region of ground water salinity have been mapped inland in Mississippi (Figure 48). These three zones are within all or part of 25 of Mississippi's 82 counties. There are also two small zones of shallow saline water, one in Washington County south of Greenville and one in Warren County starting just north of Vicksburg and extending southward. Neither of these two areas have caused any problems. Both the U.S. Geological Survey and the Mississippi Bureau of Environmental Health-Water Supply Division indicate that salinity problems in Mississippi are primarily related to oil and gas field activities (especially brine disposal).

Missouri

Salt-water intrusion is a major problem in the entire northwestern one third of Missouri (Figure 49). The shallow saline ground water which underlies the region is the source for the intrusion. Since the intrusion area (mapped in 1983) extends past the shallow saline area (mapped in 1965), it is apparent that the salt-water interface is moving southward. Ground water management efforts in Missouri should include methods which would prevent further migration of the saline/fresh water interface. Since the potential for upconing from overdrafting has been documented in Missouri, special efforts should be made to regulate ground water withdrawals, especially in the two regions in northern Missouri where shallow ground water salinity exceeds 3000 mg/l TDS. These two regions include Clay, Jackson, Ray, Carroll, Saline, Chariton, Howard, Randolph, Shelby, Monroe, Marion, Ralls, Harrison, Grundy, Livingston, and Linn Counties.

Montana

There are large areas of shallow moderately saline ground water in Montana. Occasionally, leaky wells have allowed salt-water intrusion to occur, and brine disposal associated with oil and gas field activities has caused the deterioration of ground water quality. By far, however, the major problem in Montana is dry land farming which results in 'saline seep'. This phenomenon occurs as irrigation water percolates through croplands, leaching salts from the soil. The ground water, which has increased in salinity, then tends to move laterally, surfacing some distance away, and then deposits the salts. An extremely large area in northeastern Montana has a high potential for saline seep (Figure 50).

Nebraska

Although a telephone survey indicated that there are some localized

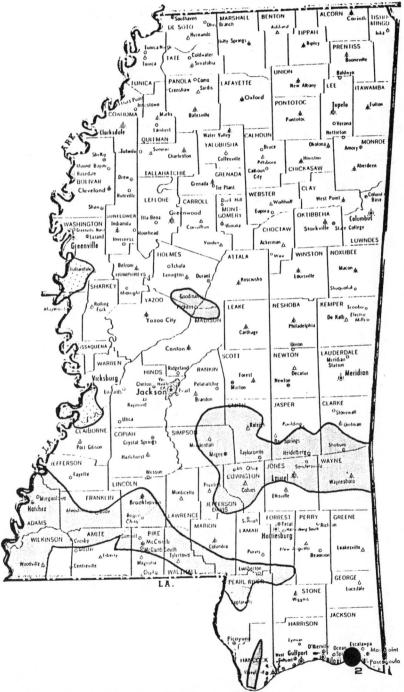

Figure 48:   Salt-water Intrusion in Mississippi

1. Miller, et al., 1977.   Salt domes throughout state.

* 2. Task Committee on Saltwater Intrusion, 1969, and Newport, 1977. Sea-water intrusion.

3. Todd, 1983.   Saline water resources.

4. Todd, 1983.   Saline water overlying fresh water.

5. Wasson, 1980.   Natural salinity.

T. Problem identified via telephone contact.

Figure 48:   (Continued)

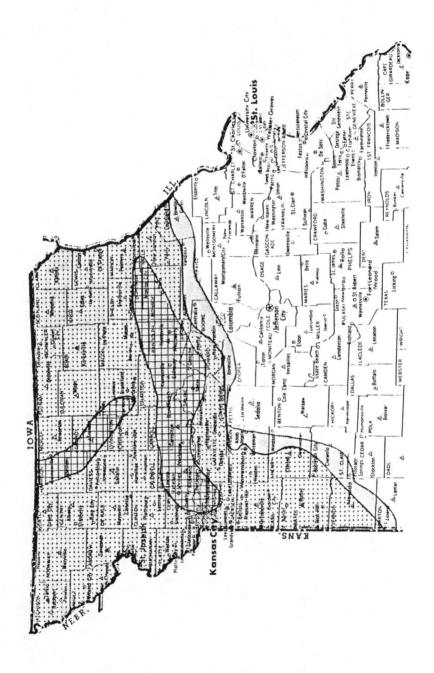

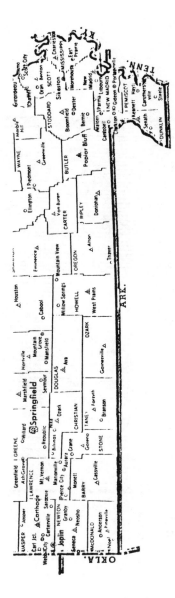

Figure 49: Salt-water Intrusion in Missouri

1. Carpenter and Darr, 1978 (AB-49). Natural salinity in Bates, Barton, and Vernon Counties.

2. Newport, 1977. Potential upconing from overdrafting.

3. Todd, 1983. Saline aquifer locations, well yields, and maximum Total Dissolved Solids.

T. Problem identified via telephone contact.

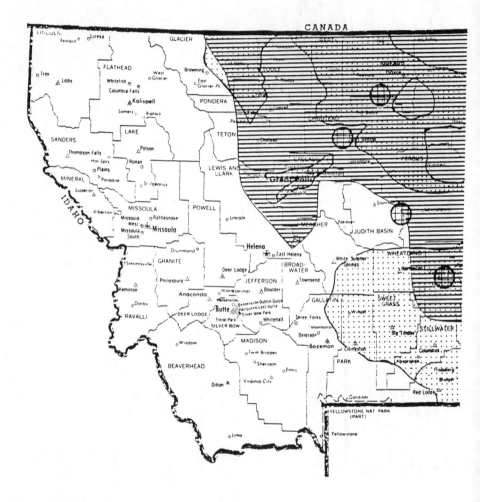

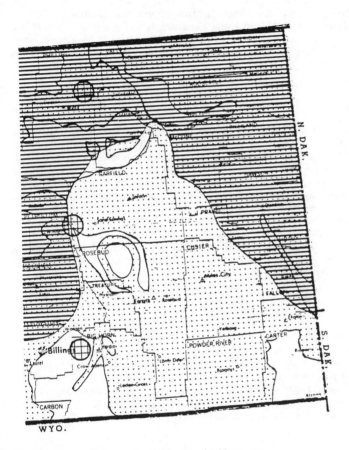

Figure 50:   Salt-water Intrusion in Montana

Bahls and Miller, 1975.   Area of potential saline seep
development.

1. Task Committee on Saltwater Intrusion, 1969.   Brine disposal and
   leaky wells in saline formation.

2. Todd, 1983.   Saline ground water resources.

problems with salt-water intrusion, Nebraska is considered one of eight states in the conterminous United States not currently experiencing problems. The areas which have been indicated as locations of intrusion are Saunders, Lancaster, and Saline Counties. Since shallow saline water exists in these counties, as well as in Deuel, Keith, Lincoln, Dundy, Dawson, Buffalo, Hall, Gage, Johnson, Otoe, Cass, Washington, Burt, Santon, Cuming, Wayne, Thurston, Dalkota, Dixon, Cedar, and Knox Counties (see Figure 51), there is a potential for localized upconing in many areas in Nebraska. However, with careful planning of well fields, Nebraska should continue to have few problems with salt-water intrusion.

Nevada

Although salt-water intrusion per se has not been documented, increasing ground water salinity in Nevada has been observed. With an average annual precipitation of about 9 inches (the lowest in the conterminous United States), Nevada relies heavily on ground water. The practice of irrigation has increased the salinity of both soil and water in areas of dry land farming. Increasing demands for irrigation water are likely to cause not only increasing soil and water salinity but also an increase in ground water withdrawal rates. Since there are numerous pockets of saline ground water in Nevada (especially the western portion--see Figure 52), an increase in ground water withdrawal may induce salt-water intrusion from the saline pockets. Churchill and Pershing Counties, especially, should be cautious since shallow ground water occurs which has a salinity with over 3000 mg/l TDS.

New Hampshire

Sea-water intrusion in Rockingham County has been mentioned in several reports (Figure 53). Overpumping of ground water resources near the coast of New Hampshire has caused lateral intrusion of Atlantic sea water. However, other types of salt-water intrusion have not been documented; and since New Hampshire has a relatively small coast, sea-water intrusion should be controllable. Once control strategies have been implemented, New Hampshire should not experience any additional salt-water intrusion.

New Jersey

Salt-water intrusion is well documented and considered a major problem in New Jersey (Figure 54). Salem, Cumberland, and Cape May Counties along the Delaware Bay have experienced serious sea-water intrusion problems. In fact, Gloucester and Atlantic Counties, which are 20 miles or more inland from Delaware Bay, are experiencing sea-water intrusion. On the Atlantic Coast, Middlesex, Monmouth, Ocean, Atlantic, and Cape May Counties have experienced sea-water intrusion. Numerous reports have cited overpumping, leaky casings, and dredging as the causes of intrusion in New Jersey. As long as water demand in New Jersey stresses the ground water resources of the state, sea-water intrusion can be expected to continue.

New Mexico

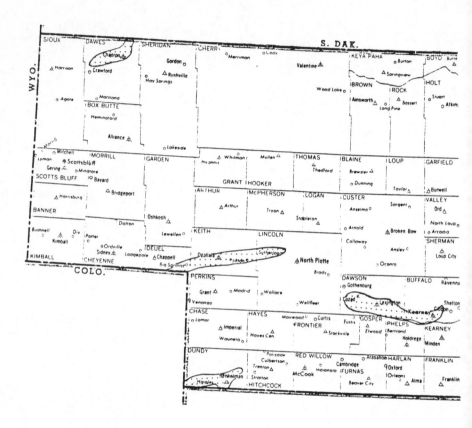

Figure 51:   Salt-water Intrusion in Nebraska

1. Task   Committee   on   Saltwater   Intrusion,   1969.   Potential
   upconing.

2. Todd, 1983. Saline ground water resources.

* T. Problem identified via telephone contact.

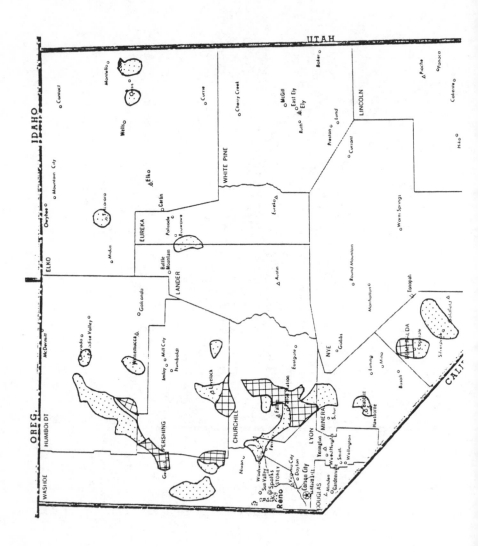

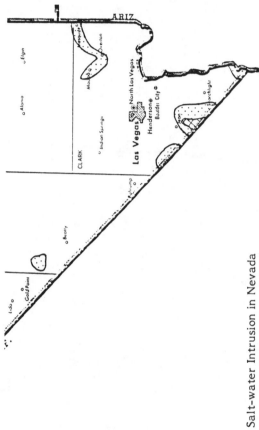

Figure 52:   Salt-water Intrusion in Nevada

1. U.S. Water Resources Council, 1981. Contamination of agricultural runoff.

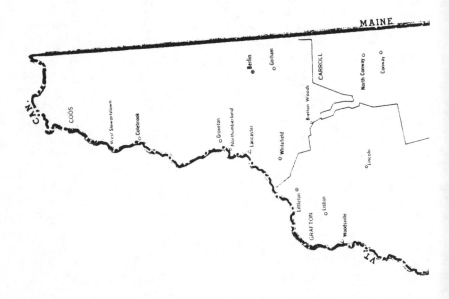

Figure 53:   Salt-water Intrusion in New Hampshire

* 1. Miller, DeLuca, and Tessier, 1974. Lateral intrusion of tidal water.

* 2. Newport, 1977. Lateral intrusion of tidal water.

* 3. Task Committee on Saltwater Intrusion, 1969. Lateral intrusion of tidal water.

* 4. Todd, 1983. Lateral intrusion of tidal water.

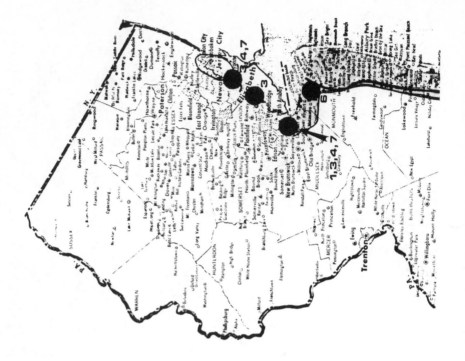

Figure 54:   Salt-water Intrusion in New Jersey

* 1. Appel, 1962. Salt-water intrusion due to over-
      pumping.

* 2. Gill, 1962. Salt-water intrusion due to overpumping.

* 3. Miller, DeLuca, and Tessier, 1974. Salt-water
      intrusion due to overpumping and leaky casings.

* 4. Newport, 1977. Salt-water intrusion due to overpumping
      and dredging.

* 5. Rosenau, et al., 1969 (AB-51). Salt-water intrusion
      due to overpumping.

* 6. Schaefer and Walker, 1981 (AB-35). Salt-water intrusion due to
      overpumping.

* 7. Task Committee on Saltwater Intrusion, 1969. Salt-water
      intrusion due to overpumping and dredging.

* 8. U.S. Water Resources Council, 1981. Natural salt water in the
      coastal plain.

The majority of New Mexico is underlain by saline ground water. In fact, only three counties are completely free of shallow saline ground water: Grant, Hidalgo, and Taos (see Figure 55). A large area of highly saline ground water exists along the eastern border of New Mexico and at least two cases of salt-water intrusion appear to be related to this zone. Both leaking wells and upconing have been documented in Lea County and Chaves County, respectively. Since most of New Mexico is considered arid, demands for usage of ground water resources can be expected to increase. These increasing water demands can potentially induce salt-water intrusion from the zone of shallow saline ground water. Furthermore, contamination from agricultural runoff has been cited as a problem which increases the potential for ground water salinity problems.

New York

Sea-water intrusion has been occurring in New York for at least 20 years. Kings, Queens, Nassau, and Suffolk Counties (all on Long Island) have experienced serious threats to their ground water resources from overpumping and subsequent sea-water intrusion. However, using treated waste-water and storm water, a network of artificial recharge wells near the coast has created a hydraulic barrier to the intruding sea water. Although the threat of sea water intrusion has been reduced in most places, some areas such as Montauk Point (eastern Long Island) are still faced with intrusion. Near Buffalo, in Erie County, salt-water intrusion has also been documented. The intrusion is associated with overpumping near shallow saline ground water. The shallow saline ground water underlies a large portion of New York (see Figure 56); and if other large metropolitan areas such as Rochester, Syracuse, and Albany begin to develop ground water well fields as Buffalo has, more salt-water intrusion may occur.

North Carolina

A zone of shallow moderately saline ground water occurs in North Carolina along the Atlantic Ocean (Figure 57). Associated with this zone, salt-water intrusion has been documented in Carteret, Pamlico, Beaufort, Hyde, Dare, and Tyrrell Counties. One report has cited salt-water intrusion problems near the cities of New Bern and Wilmington, while the North Carolina Department of Natural Resources and Community Development-Ground Water Section indicates that sea-water intrusion is occurring along the outer banks in areas such as Nags Head and Kill Devil. Other than these areas along the coast, North Carolina should not experience salt-water intrusion since there are no other areas of shallow saline ground water.

North Dakota

The entire State of North Dakota can expect to experience increasing salt-water intrusion problems unless significant efforts to control the problems are implemented. As shown in Figure 58, shallow saline ground water completely underlies the state, and potential saline seep has been documented for nearly the entire state. Saline seep problems are increased due to artesian pressure on deep saline aquifers. In some areas (especially the Red River Valley) the piezometric head of deep saline aquifers are at higher elevations

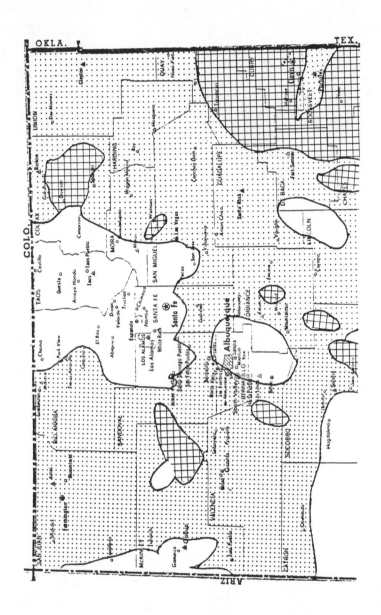

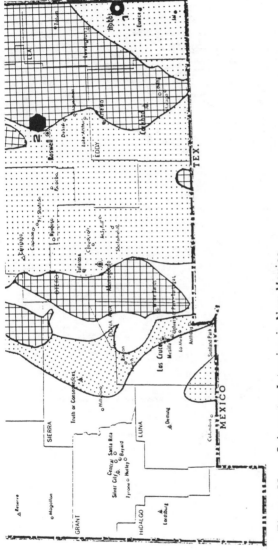

Figure 55:   Salt-water Intrusion in New Mexico

* 1. Anonymous, 1985. Leaking wells.

* 2. Mattox, 1970. Upconing.

  3. Newport, 1977. Upconing.

  4. U.S. Water Resources Council, 1981. Natural salinity, northern
     part; upconing and contamination from agricultural runoff.

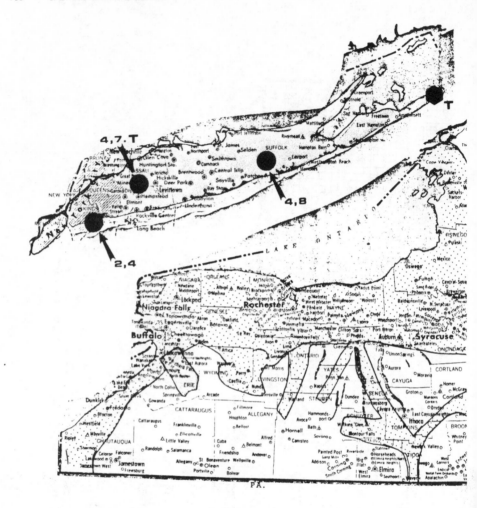

Figure 56:   Salt-water Intrusion in New York

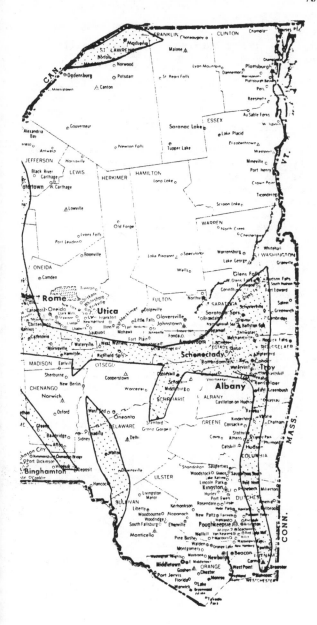

1. Anderson and Berkebile, 1976 (AB-19).   Leaky wells in Southeastern Long Island.

* 2. Buxton, et al., 1981 (AB-20).   Salt-water intrusion in Kings and Queens Counties, Long Island.

3. Collins, 1972 (AB-87).   Salt-water intrusion in Long Island.

* 4. Miller, DeLuca, and Tessier, 1974.   Upconing in Western part of state; salt-water intrusion from lowering of water table in Kings County, Long Island; and salt-water intrusion due to pumping, leaky well casings, and dredging in Queens, Nassau, and Suffolk Counties, Long Island.

5. Montanari and Waterman, 1968 (AB-72).   Sea-water intrusion on Long Island.

6. Newport, 1977.   Sea-water intrusion on Long Island.

* 7. Peters and Rose, 1968 (AB-75).   Salt-water intrusion in Nassau County, Long Island.

* 8. Soren, 1978 (AB-39).   Salt-water intrusion on Long Island.

9. Task Committee on Saltwater Intrusion, 1969.   Sea-water intrusion on Long Island.

10. Todd, 1983.   Sea-water intrusion on Long Island.

11. U.S. Water Resources Council, 1981.   Sea-water intrusion due to overpumping on Long Island.

* T. Problem identified via telephone contact.

Figure 56:   (Continued)

Figure 57:   Salt-water Intrusion in North Carolina

  1. Kritz, 1972. Pumping pattern.

* 2. Task Committee on Saltwater Intrusion, 1969, and Newport, 1977.
     Sea-water intrusion.

  3. Todd, 1983. Saline water overlying fresh water.

* T. Problem identified via telephone contact.

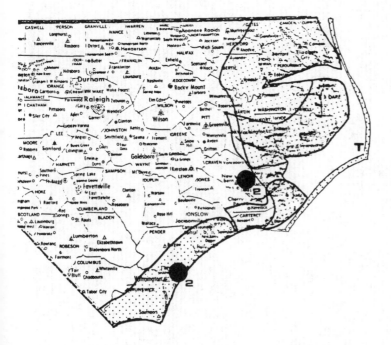

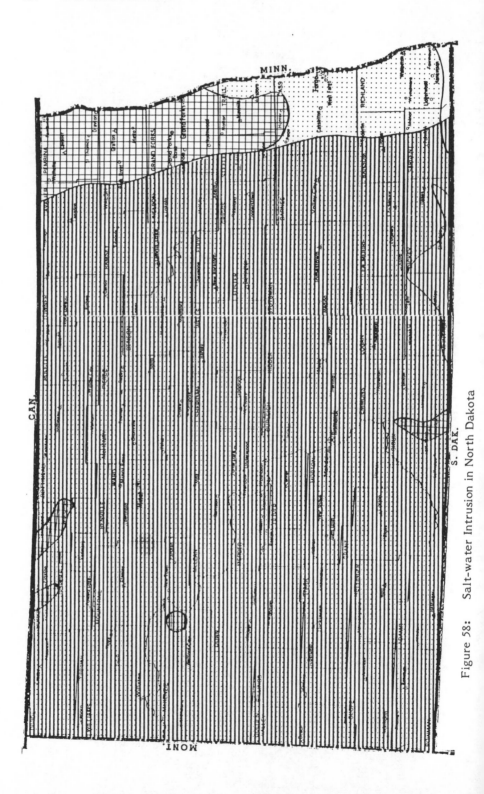

Figure 58:   Salt-water Intrusion in North Dakota

Bahls and Miller, 1975. Area of potential saline seep development.

1. Task Committee on Saltwater Intrusion, 1969. Upconing (Red River Valley).

2. Todd, 1983. Saline ground water resources.

3. U.S. Water Resources Council, 1981. Natural salinity problem.

T. Problem identified via telephone contact.

Figure 58: (Continued)

than the ground surface, resulting in discharges of saline water to the surface. Upconing has also been documented in the Red River Valley. Salt-water intrusion problems are increased by oil and gas field brines associated with oil and gas production in the state.

Ohio

Shallow, moderately saline ground water occurs throughout the entire state (Figure 59). One pocket of natural salinity very near the surface has also been documented in the northeastern corner of the state in Ashtabula, Lake, and Geauga Counties. Although the potential for upconing and lateral intrusion are constant threats in Ohio, to date they have not been the major types of salt-water intrusion. Both the Department of Natural Resources--Water Division, and the Water Resources Center (Ohio State University) indicate that brine disposal from oil and gas field activities is the major problem in Ohio. However, if ground water resources are more fully developed in the state, salt-water intrusion problems will probably increase.

Oklahoma

Almost all of Oklahoma is underlain by shallow, moderately saline ground water. There is also a pocket of highly saline ground water in the southwestern corner of the state (Beckham, Greer, Garmon, and Jackson Counties), and one along the northern border (Alfalfa County). Furthermore, as pointed out in Figure 60, additional problems of natural salinity and oil field brines have been documented in the Cimarron Terrace (from Woods County southeast to Logan County). The Water Resources Division of the U.S. Geological Survey has indicated that there is extensive occurrence of salt-water intrusion in Oklahoma, especially in areas of overdevelopment, but the problem has been given little attention. The Oklahoma Water Resources Board--Ground Water Division--indicates that overdevelopment of the Garber-Wellington Aquifer (in central Oklahoma), coupled with many poorly constructed wells, has resulted in many local areas of salt-water intrusion.

Oregon

Sea-water intrusion, upconing, lateral intrusion, and underground injection have all been documented in Oregon. The majority of problems with salt-water intrusion have occurred in the northwestern corner of the state (see Figure 61). Sea-water intrusion has been documented in Clatsop, Tillamook, and Coos Counties. Upconing and/or lateral intrusion has been documented near Portland in Multnomah County and near Salem in Yamhill, Marion, and Polk Counties. The U.S. Geological Survey has pointed out that salt-water intrusion problems also exist around Eugene (Lane County), North Bend (Coos County), and in the Willamette Valley in the West Cascades. Shallow saline ground water occurs in small pockets throughout the state, but areas other than those around Portland have not produced many problems.

Pennsylvania

Although almost the entire western half of the state is underlain by

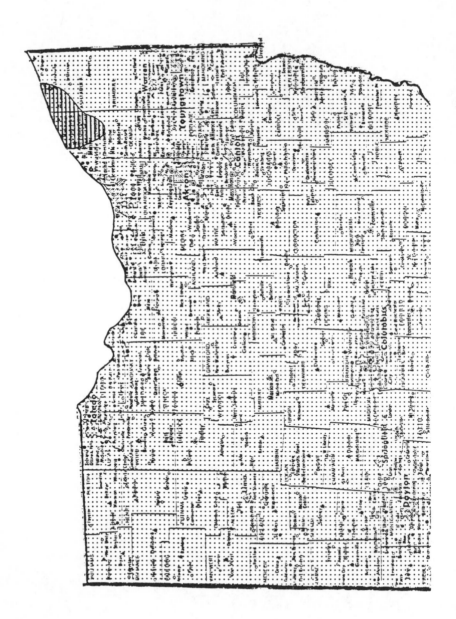

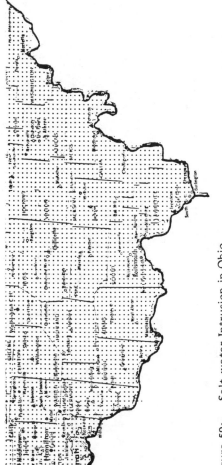

Figure 59:   Salt-water Intrusion in Ohio

Norris, 1978. Natural salinity.

1. Task Committee on Saltwater Intrusion, 1969, and Newport, 1977.
   Oil field brines.

T. Problem identified via telephone contact.

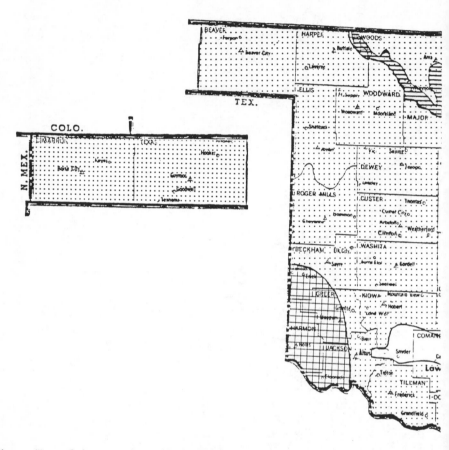

Figure 60:   Salt-water Intrusion in Oklahoma

1. Cities Service Oil Co. vs. Merritt, 1958 (AB-54).   Pollution
   during drilling.

2. Newport, 1977. Oil field brines.

3.   Oklahoma Water Resources Board, 1975.   Natural
      salinity and oil field brines in the Cimarron Terrace.

* 4. U.S. Water Resources Council, 1981.   Natural salinity, Caddo
   County and along the Salt Fork of the Arkansas River.

T. Problem identified via telephone contact.

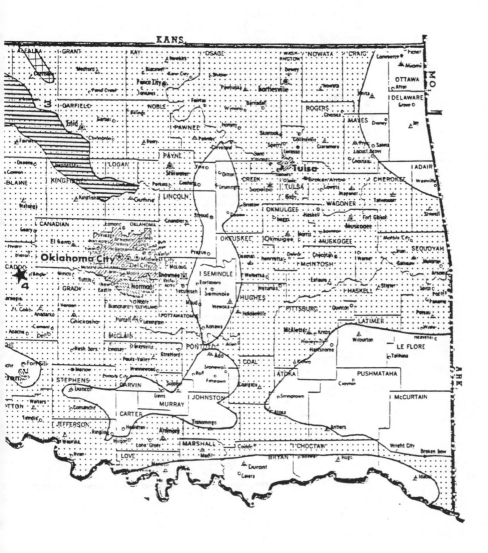

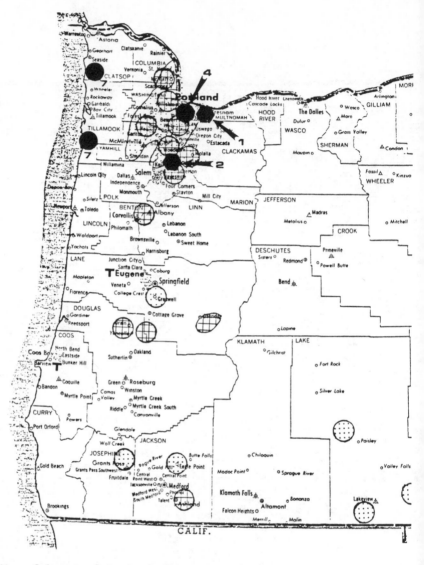

Figure 61:   Salt-water Intrusion in Oregon

* 1. Brown, 1963. Upconing of saline water (Portland).

* 2. Hampton, 1972. Upconing of saline water (Molalla-Salem Slope Area, North Williamette Valley).

  3. Miller, 1980. Well disposal practice.

* 4. Price, Hart, and Foxworthy, 1962. Pumpage of wells and lowering of the hydrostatic head caused saline water intrusion (Portland).

  5. Task Committee on Saltwater Intrusion, 1969.

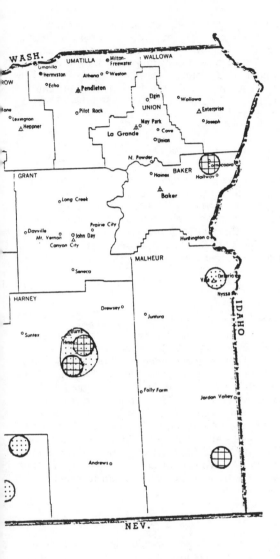

6. Todd, 1983. Sea-water intrusion. (Coasts).

* 7. Schlicker, et al., 1970.   Sea-water intrusion (Tillamook and Clatsop Counties).

8. U.S. Water Resources Council, 1981.   Localized salt-water intrusion.

9. Van der Leeden, Cerillo, and Miller, 1975.   Sea-water intrusion.

* T. Problem identified via telephone contact.

Figure 61:   (Continued)

shallow moderately saline ground water, most documented salt-water intrusion problems have occurred in the eastern part of the state, near Philadelphia (Figure 62). These problems have been associated with overpumping near, and dredging in, the tidal river areas. There have been, however, a few cases of upconing reported in the western part of the state. Although there are a few incidents of salt-water intrusion occurring in Pennsylvania, the Bureau of Commerce and Environmental Control--Division of Water Supply--indicates that salt-water intrusion is not a problem.

Rhode Island

Several reports have documented the occurrence of sea-water intrusion in Rhode Island (Figure 63). The intrusion has occurred in Bristol and Kent Counties near Warwick and in Providence County near Providence. Overpumping near the tidal river estuaries has been cited as the cause for the sea-water intrusion. The remainder of the state, however, has not experienced any problems with salt-water intrusion and should not expect problems away from the coast since there are no pockets of shallow saline ground water.

South Carolina

Salt-water intrusion has been documented in several areas of South Carolina, and it can be considered a major problem in several of the locations. Almost the entire South Carolina Atlantic coastline has experienced sea-water intrusion (Figure 64), and three major studies have addressed the problems of sea-water intrusion around Charleston, Burford, and Horry Counties. In addition to sea-water intrusion, South Carolina also faces intrusion problems associated with a zone of salt-water which extends from the southwest to the northeast in the center of the state. This zone is associated with an unusual geological formation where an abundance of salt water is located. The entire zone has experienced various degrees of salt-water intrusion. Both upconing and downconing have occurred from the zone since fresh ground water occurs both above and below the salt water. As ground water resources are further developed in this area, more salt-water intrusion can be expected to occur.

South Dakota

Although there are two large zones of shallow, moderately saline ground water and two small zones of shallow highly saline ground water in South Dakota, the major salt-water problem is associated with a very large zone of potential saline seep (Figure 65). As in North Dakota and Montana, this large area of potential saline seep is created by salt-water aquifers under artesian pressure producing the possibility of salt-water discharges to the surface or into fresh water aquifers through hydrologic connections. However, upconing has been reported in one area not associated with the zone of potential saline seep or shallow saline ground water--the Black Hills in the southwestern corner of the state.

Tennessee

Tennessee is one of the eight states in the conterminous United States

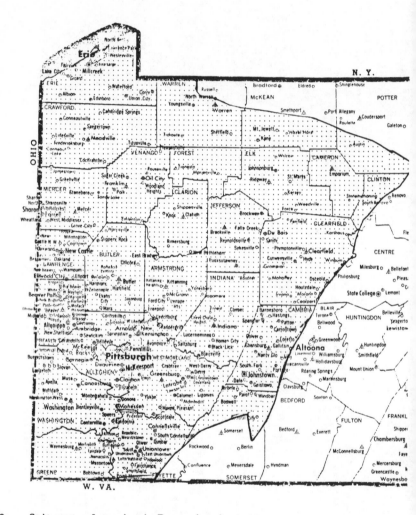

Figure 62:   Salt-water Intrusion in Pennsylvania

* 1. Miller, DeLuca, and Tessier, 1974. Upconing in western part of
     state and lateral intrusion of salt water from tidal river due
     to overpumping, leaky casings, and dredging.

* 2. Newport, 1977.   Lateral intrusion of salt water due to
     overpumping and dredging.

* 3. Task Committee on Saltwater Intrusion, 1969. Lateral intrusion
     of salt water due to overpumping and dredging.

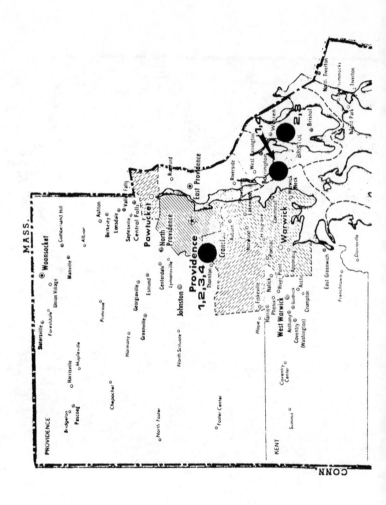

Figure 63:   Salt-water Intrusion in Rhode Island

* 1. Miller, DeLuca, and Tessier, 1974. Sea-water intrusion from tidal river estuaries due to overpumping.

* 2. Newport, 1977. Sea-water intrusion from tidal river estuaries due to overpumping.

* 3. Task Committee on Saltwater Intrusion, 1969. Sea-water intrusion from tidal river estuaries due to overpumping.

* 4. Todd, 1983. Sea-water intrusion from tidal river estuaries due to overpumping.

T. Problem identified via telephone contact.

Figure 64:   Salt-water Intrusion in South Carolina

* 1. Hardee and McFadden, 1982. Sea-water intrusion.

  2. Miller, et al., 1977. Salt-water intrusion.

* 3. Science and Public Policy Program, 1983. Salt-water intrusion.

* 4. Task Committee on Saltwater Intrusion, 1969, and Newport, 1977.
     Upconing and downconing. Sea-water intrusion.

  5. Todd, 1983. Saline water overlying fresh water.

* T. Problem identified via telephone contact.

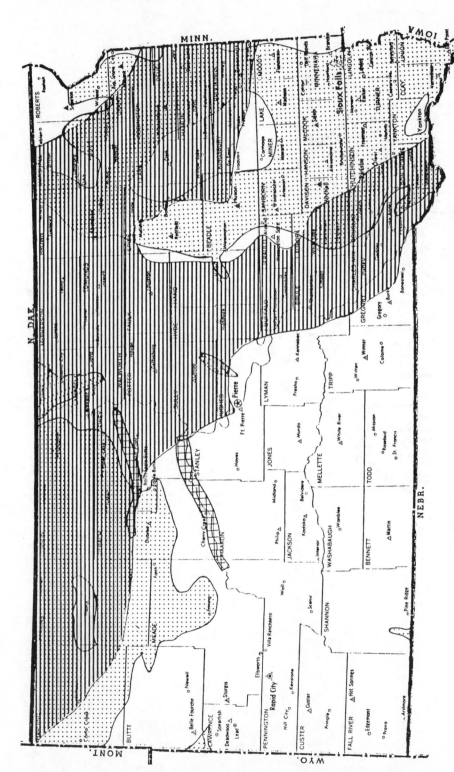

Figure 65:   Salt-water Intrusion in South Dakota

▓ Bahls and Miller, 1975. Area of potential saline seep development.

1. Meyer, 1981. Ground water pollution sources.

2. Meyer and Barwell, 1983. Ground water resources.

3. Task Committee on Saltwater Intrusion, 1969. Upconing of saline water (Black Hills).

4. Todd, 1983. Saline ground water resources.

T. Problem identified via telephone contact.

Figure 65:   (Continued)

which has not reported any problems with salt-water intrusion. However, there is a large zone of shallow, highly saline ground water surrounded by an even larger zone of shallow, moderately saline ground water located in the center of the state (Figure 66). Although problems have not yet been documented, Nashville is situated directly over the high salinity zone. If water demand induces concentrated development of ground water resources in the Nashville area, problems are likely to occur. Water resource planners should exercise caution when developing ground water well fields in central Tennessee.

Texas

There have been numerous incidents of salt-water intrusion in Texas. Along the Gulf Coast, sea-water intrusion has been documented in Galveston, Texas City, and Corpus Christi. Southeast of Houston, overpumping of ground water has lead to large areas of subsidence which subsequently has resulted in sea-water intrusion. Several sea-water intrusion models have been developed based on the problems experienced near Houston. Additional, smaller scale, localized sea-water intrusion problems have been encountered all along the coast. The U.S. Geological Survey--Water Resources Division--has helped establish various programs for controlling the sea-water intrusion. Oil and gas field activities have also been blamed for salt-water contamination in Texas. Drilling, evaporation pits, and brine injection have all been cited as sources. Upconing has been documented in El Paso, Dimmit, Kinney, and Hays Counties.

In addition to salt-water intrusion problems already documented in Texas, the state must contend with large areas of shallow saline ground water (Figure 67). Almost the entire northwestern part of the state (the panhandle) is underlain by shallow, moderately saline ground water. Since this is the area where the Ogallala Aquifer is rapidly being depleted, it can be expected that salt-water intrusion will occur unless control measures are soon implemented. There are also other areas of shallow saline ground water. Starting around the western edge of Uvalde County (west of San Antonio) and stretching northeast past Austin to Williamson County is a narrow zone of saline water. This area, known in Texas as the 'Bad Water Zone', borders the Edwards Aquifer, a major source of water for south-central Texas. Plans are currently underway to develop surface water supplies for the area to prevent further depletion of the Edwards which could otherwise induce salt-water intrusion from the Bad Water Zone.

Utah

Although there have been few documented cases of salt-water intrusion in Utah, the Utah Water Research Laboratory of Utah State University has indicated that salinity problems do occur in the eastern portion of the state at the upper end of the Colorado River where marine shales outcrop. The potential for upconing/lateral intrusion near the Great Salt Lake has prompted careful planning for ground water development in the area. Also, there are several fragmented zones of moderately and highly saline shallow saline ground water throughout the state (see Figure 68). Since many of the highly saline zones are extensive, ground water resource development in Utah should proceed with caution.

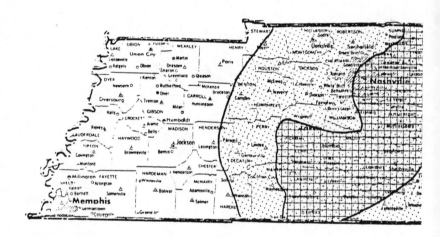

Figure 66:   Salt-water Intrusion in Tennessee

1. Task Committee on Saltwater Intrusion, 1969, and Newport, 1977.
No known examples.

T. Problem identified via telephone contact.

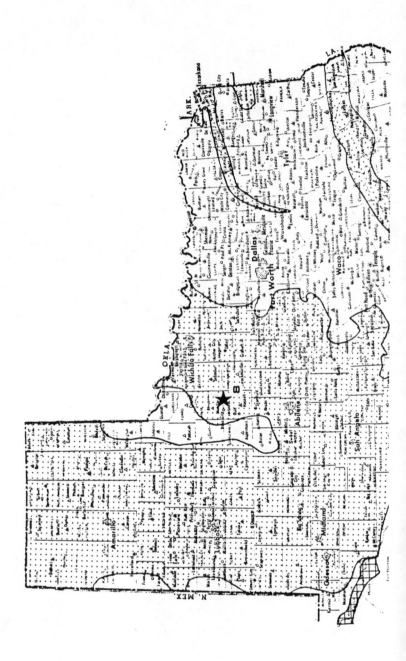

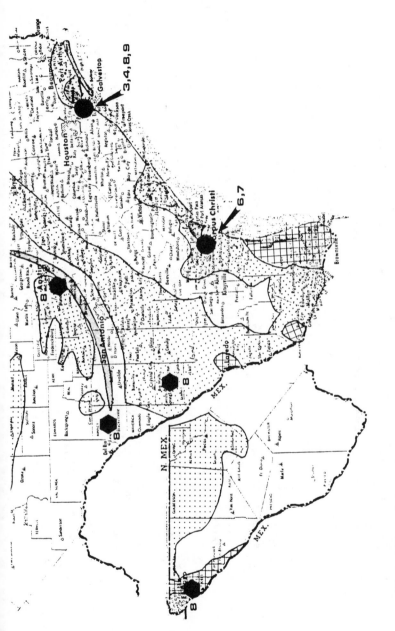

Figure 67: Salt-water Intrusion in Texas

1. Conselman, 1970. Natural salinity, West Texas.

2. General Crude Oil vs. Aiken, 1960 (AB-56). Pollution from oil well evaporation tanks.

* 3. Jorgenson, 1981 (AB-116). Sea-water intrusion models, Houston.

* 4. Newport, 1977. Sea-water intrusion, Galveston and Texas City.

5. Pickens vs. Harrison, 1952 (AB-58). Pollution from oil well drilling.

* 6. Shafer, 1968 (AB-60). Sea-water intrusion, Nueces and San Patricio Counties.

* 7. Shafer, 1970 (AB-37). Sea-water intrusion, Aransas County.

* 8. Texas Water Development Board, 1977. Sea-water intrusion, Galveston; upconing, El Paso, Dimmit, Kinney, and Hays Counties; natural salinity, Haskell, Jones, Knox, and Baylor Counties.

* 9. Todd, 1983. Sea-water intrusion along coast and in Harris County.

10. U.S. Water Resources Council, 1981. Localized sea-water intrusion along the coast; contamination from agricultural runoff in the Rio Grande Region.

T. Problem identified via telephone contact.

Figure 67:   (Continued)

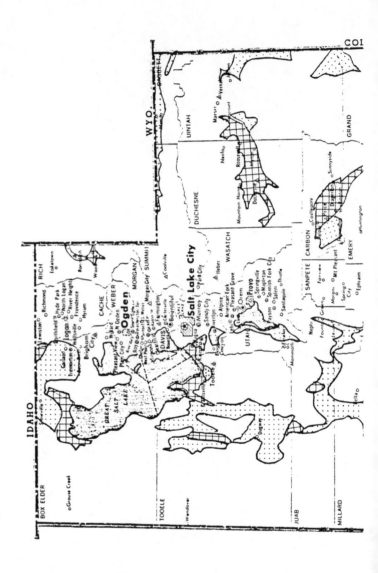

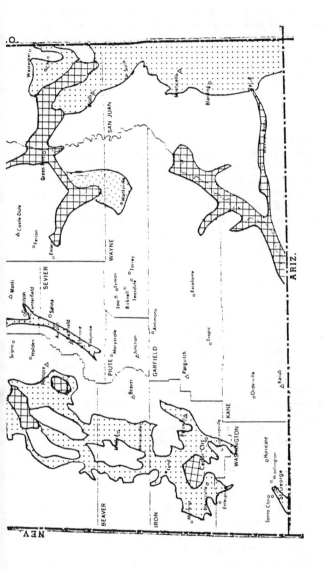

Figure 68:   Salt-water Intrusion in Utah

1. Newport, 1977. Potential upconing problem in the Great Salt Lake area and the western part of the state.

2. U.S. Water Resources Council, 1981. Natural salinity.

T. Problem identified via telephone contact.

Vermont

The Division of Environmental Health in Vermont's Department of Health has indicated that there are some small areas where naturally occurring saline ground water exists. However, it is estimated that less than 50 wells in the state have been affected by the problem. In fact, the areas of shallow saline ground water are so small that they are not mapped (see Figure 69). Therefore, Vermont is considered one of the eight states in the conterminous United States that has no salt-water intrusion problems.

Virginia

Sea-water intrusion has been documented in Virginia in Northhampton and Isle of Wight Counties. These cases of sea-water intrusion are associated with overpumping and a zone of shallow saline ground water connected to the Atlantic Ocean. Virginia's Water Council Board operates a 15-well monitoring network to measure chloride concentration as a part of a control program. Other than the problems along the coast, Virginia has not experienced any other salt-water intrusion problems. In fact, there is only one other pocket of shallow saline ground water in the state. As indicated in Figure 70, it is a very small area in Giles County underlying Pearisburg and Narrows. To date, there have not been any problems in Virginia associated with this zone of saline water.

Washington

Several cases of sea-water intrusion associated with Puget Sound have been reported. At the northern end of the Sound, Island County has experienced problems, while at the southern end, sea-water intrusion has occurred in many areas around Tacoma and Olympia and in Kitsap County. Along the Pacific Coast, sea-water intrusion has been documented in northern Jefferson County. Finally, away from the coast, salt-water intrusion has been documented in Grant County, where intrusion from a saline lake was induced by overpumping. There are only a few pockets of shallow saline ground water in Washington (see Figure 71); and although some of them contain highly saline water (over 3000 mg/l TDS), there have not been associated intrusion problems.

West Virginia

West Virginia is considered one of the eight states in the conterminous United States which is not experiencing salt-water intrusion problems. However, the potential for problems has been recognized in at least two reports as well as in a telephone survey. One study investigated the importance of proper pumping rates to prevent intrusion from saline water which underlies the fresh ground water in Calhoun, Mason, Wayne, and Boone Counties. However, as indicated in Figure 72, over half of the state is underlain by shallow, moderately saline ground water, and proper planning for ground water withdrawals should help prevent future salt-water intrusion problems.

Wisconsin

Salt-water intrusion is a minor problem in Wisconsin, although some cases

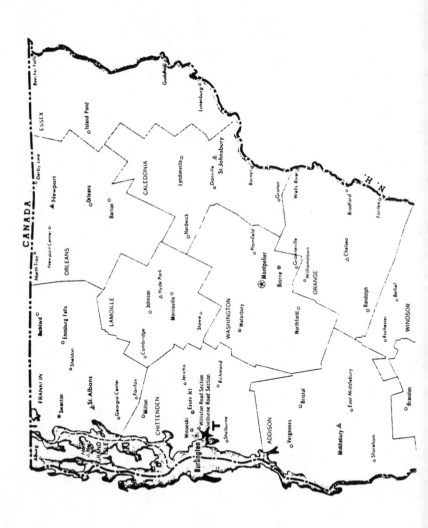

Figure 69:   Salt-water Intrusion in Vermont

* T. Problem identified via telephone contact.

Figure 70:   Salt-water Intrusion in Virginia

1. Larson, 1981. Salt-water intrusion due to overpumping.

* 2. Newport, 1977. Sea-water intrusion.

* 3. Task   Committee   on   Saltwater   Intrusion,   1969.   Sea-water
   intrusion.

* 4. U.S. Water Resources Council, 1981. Salt-water intrusion due to
   overpumping.

   T. Problem identified via telephone contact.

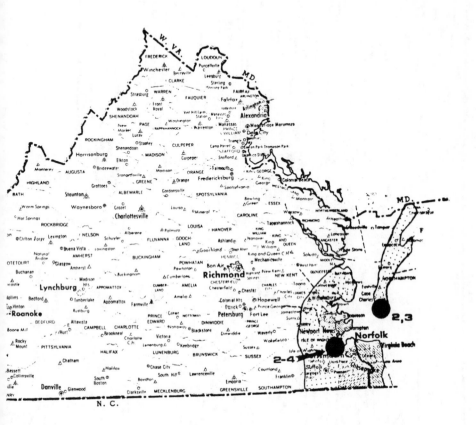

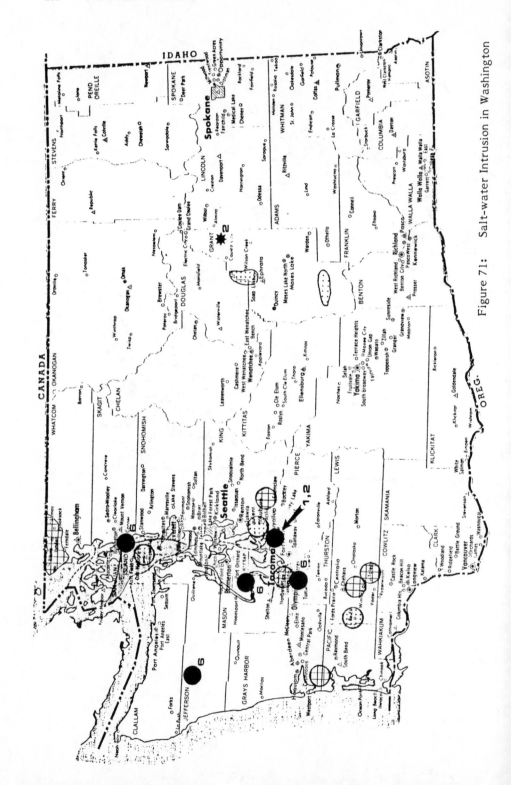

Figure 71:   Salt-water Intrusion in Washington

* 1. Kimmel, 1963. Sea-water intrusion (Hyada Park at Tacoma).

* 2. Task Committee on Saltwater Intrusion, 1969. Sea-water intrusion, Tacoma Area, and intrusion from saline lake, Grant County.

3. Todd, 1983. Sea-water intrusion (Puget Sound).

4. U.S. Water Resources Council, 1981. Localized salt-water intrusion.

5. Van der Leeden, Cerillo, and Miller, 1975. Sea-water intrusion.

* 6. Walters, 1971 (AB-44). Sea-water intrusion, Puget Sound, Olympia at Tacoma, Northern Kitsap and Northern Jefferson Counties, and in Island County.

Figure 71: (Continued)

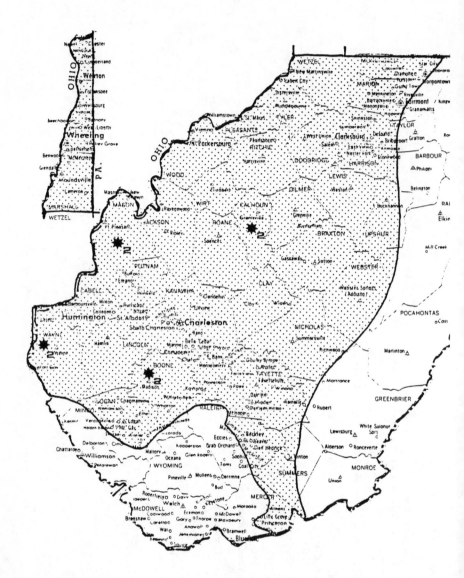

Figure 72:   Salt-water Intrusion in West Virginia

1. Task Committee on Saltwater Intrusion, 1969, and Newport, 1977.
   No known examples; however, potential exists.

* 2. Wilmoth, 1975 (AB-53). Pumping patterns.

T. Problem identified via telephone contact.

have been documented. In Manitowoc County, on the shore of Lake Michigan, salt-water intrusion was reported in the early 1960's. Since that time, few problems have occurred. There are, however, several pockets of shallow saline ground water in the state (Figure 73). In northern Wisconsin, all along the Lake Superior shoreline, moderately saline ground water exists at depths of less than 500 feet. Douglas, Bayfield, Ashland, and Iron Counties are affected. Along the Lake Michigan shoreline, Marineette, Kewaunee, Manitowoc, and Milwaukee Counties overlie shallow, moderately saline water. Sheboygan and Ozaukee Counties both have pockets of highly saline ground water (TDS greater than 3000 mg/l). Slightly away from the Lake Michigan shore is a somewhat larger pocket of shallow, moderately saline water reaching from Brown County to Fon Du Lac County. Florence County, along the Michigan border, has a small pocket of highly saline ground water. Ground water development should be carefully planned in these areas, or salt-water intrusion may occur in the future.

Wyoming

Figure 74 indicates the areas of shallow saline ground water in Wyoming; 18 of Wyoming's 23 counties are underlain in whole or in part by numerous shallow moderately saline aquifers. Three small pockets of highly saline ground water exist, two in Lincoln County and one in Johnson County. Although there are abundant sources of salt water in Wyoming and upconing or lateral intrusion might be expected, the major problem has been reported as resulting from oil and gas field activities. Ground water resources are not heavily relied upon in Wyoming, and salt-water intrusion will probably not increase unless this resource is more fully developed in the future. Careful planning of future ground water development and oil and gas field practices should keep salt-water intrusion to a minimum in Wyoming.

Summary

Salt-water intrusion is a common problem throughout the nation. Of the 48 contiguous states, 40 have reported incidents of intrusion. Every state along the Altantic, Pacific, and Gulf of Mexico shores have experienced sea-water intrusion. The problems are both regional and isolated, resulting in the problems being of various magnitudes in each of the 40 impacted states. While some states have reported only one or two leaking wells, for example, others indicate that salt-water intrusion is a wide-spread phenomenon throughout the state. The eight states in the contiguous states not experiencing problems all have the potential for salt-water intrusion problems due to shallow saline ground water. As pressures increase to develop ground water resources in these states, intrusion problems will likely also increase. In areas where highly saline water occurs, the risks of intrusion are greater since a smaller volume of this salinity will generate greater problems. When all factors are considered, salt-water intrusion can be considered a significant threat to the nation's ground water resources.

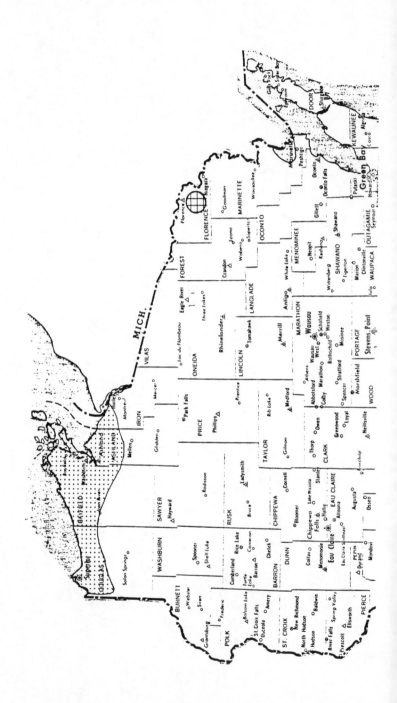

Figure 73:   Salt-water Intrusion in Wisconsin

1. Kammerer, 1981.   Ground water quality.

* 2. Ryling, 1961.   Saline ground water resources and salt-water intrusion, Eastern Manitowoc County.

3. Task Committee on Saltwater Intrusion, 1969.   Lateral salt-water intrusion.

T. Problem identified via telephone contact.

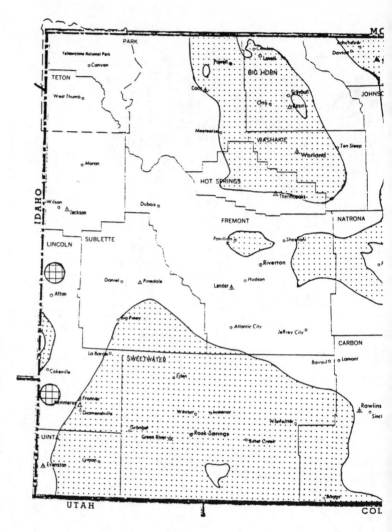

Figure 74:   Salt-water Intrusion in Wyoming

1. Task   Committee   on   Salt-water   Intrusion,   1969.   Salt-water
   intrusion by oil well activities and irrigation return flows.

2. Todd, 1983.   Saline ground water resources.

T. Problem identified via telephone contact.

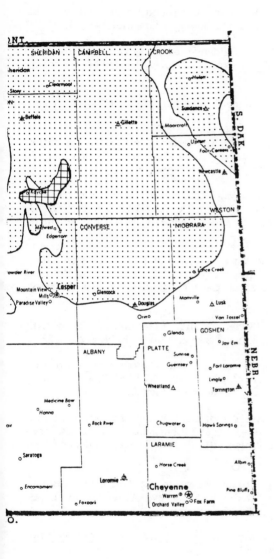

SELECTED REFERENCES

Abegglen, D.E., "The Effect of Drain Wells on the Ground Water Quality of the Snake River Plain", Idaho Bureau of Mines and Geology Pamphlet 148, 1970.

Anderson, M.P., and Berkebile, C.A., "Evidence of Salt-Water Intrusion in Southeastern Long Island", Ground Water, Vol. 14, No. 5, September-October, 1976, pp. 315-319.

Anonymous, "New Mexico Aquifer is Facing Severe Salt Water Threat", Ground Water, January-February, 1985.

Appel, C.A., "Salt-Water Encroachment into Aquifers of the Raritan Formation in the Sayreville Area, Middlesex County, New Jersey", Special Report No. 17, 1962, Division of Water Policy and Supply, Trenton, New Jersey.

Aral, M.M., Sturm, T.W., and Fulford, J.M., "Analysis of the Development of Shallow Groundwater Supplies by Pumping from Ponds", OWRT-A-089-GA(1), February, 1981, Office of Water Research and Technology, U.S. Department of the Interior, Washington, D.C.

Bahls, L.L., and Miller, M.R., "Ground Water Seepage and Its Effects on Saline Soils-Completion Report", MUJWRRC Report No. 66, October, 1975, Montana University Joint Water Resources Research Center, Montana State University, Bozeman, Montana.

Balsters, R.G., and Anderson, C., "Water Quality Effects Associated with Irrigation", Kansas Water News, Vol. 22, No. 1 and 2, Winter, 1979, pp. 14-22.

Boggess, D.H., Missimer, T.M., and O'Donnell, T.H., "Saline-Water Intrusion Related to Well Construction in Lee County, Florida", USGS/WRD/WRI-77/061, September, 1977, U.S. Geological Survey, Tallahassee, Florida.

Bookman, M., "Legal and Economic Aspects of Salt-Water Encroachment into Coastal Aquifers", Bulletin 3, October, 1968, Louisiana Water Resources Research Institute, Louisiana State University, Baton Rouge, Louisiana, pp. 159-192.

Brown, S.G., "Problems of Utilizing Ground Water in the West-Side Business District of Portland, Oregon", U.S. Geological Survey, Water-Supply Paper 1619-0, 1963.

Bruington, A.E., "Saltwater Intrusion into Aquifers", Water Resources Bulletin, Vol. 8, No. 1, February, 1972, pp. 150-160.

Bruington, A.E., "Control of Sea-Water Intrusion in Ground-Water Aquifer", Ground Water. Vol. 7, No. 3, May-June, 1969, pp. 9-14.

Burkart, M.R., "Availability and Quality of Water from the Dakota Aquifer, Northwest Iowa", Water-Supply Paper 2215, 1984, U. S. Geological Survey, Washington, D.C.

Buxton, H.T., et al., "Reconnaissance of the Ground-Water Resources of Kings and Queens Counties, New York", Open-File Report 81-1186, 1981, U.S. Geological Survey, Syosset, New York.

Carpenter, A.B., and Darr, J.M., "Relationship Between Groundwater Resources and Energy Production in Southwestern Missouri", OWRT-A-086-MO(1), April, 1978, Office of Water Research and Technology, U.S. Department of the Interior, Washington, D.C.

Cherry, R.N., "Hydrology of Clearwater-Dunedin, Florida", 1980, U.S. Geological Survey, Tampa, Florida.

Christensen, B.A., "Mathematical Methods for Analysis of Fresh Water Lenses in the Coastal Zones of the Floridan Aquifer", OWRT-A-032-FLA(1), December, 1978, Office of Water Research and Technology, U.S. Department of the Interior, Washington, D.C.

Christensen, B.A., and Evans, A.J., Jr., "A Physical Model for Prediction and Control of Salt Water Intrusion in the Floridian Aquifer", Publication No. 27, January, 1974, Water Resources Research Center, University of Florida, Gainesville, Florida.

Cities Service Oil Co. vs. Merritt (Damages for the Pollution of Subterranean Waters), 332 P2D 677-688 (OKLA. 1958).

Collins, M.A., "Ground-Surface Water Interaction in the Long Island Aquifer System", Water Resources Bulletin, Vol. 8, No. 6, December, 1972, pp. 1253-1258.

Conselman, F.B., "Our Undeveloped Subsurface Water Resources - Their Potential and Their Challenge", Proceedings of the 8th West Texas Water Conference, Texas Tech University, Lubbock, Texas, 1970, pp. 25-28.

Daniel, D.L., "Map Showing Total Dissolved Solids Content of Ground Water in Arizona", Hydrological Report No. 2, Department of Water Resources in Cooperation with the Arizona Department of Health Services, January, 1981.

Department of Water Resources, The Resources Agency, State of California, "Sea-water Intrusion in California: Inventory of Coastal Ground Water Basins", Bulletin No. 63-5, October, 1975.

Feth, J.H., et al., "Preliminary Map of the Conterminous U.S. Showing Depth to and Quality of Shallowest Ground Water Containing More Than 1000 Parts Per Million Dissolved Solids", Atlas HA-199, 1965, U.S. Geological Survey, Washington, D.C.

Florida House of Representatives, "Report of the Select Committee on Water Resources (Foreign Matter and Salt Water Intrusion in Public Waters)", Tallahassee, Florida, October, 1964, pp. 12-15.

Florida House of Representatives, "Report of the Select Committee on Water Resources (Florida's Water Problems)", Tallahassee, Florida, October, 1964.

Fairchild, R.W., and Bentley, C.B., "Saline-Water Intrusion in the Floridan Aquifer in the Fernandina Beach Area, Nassau County, Florida", USGS/WRD/WRI-77/075, September, 1977, U.S. Geological Survey, Tallahassee, Florida.

Fractor, D.T., "Property Rights Approach to Groundwater Management",

Dissertation Abstracts International, Vol. 43, No. 9, 1983, p. 3064.

Franks, B.J., and Phelps, G.G., "Estimated Drawdowns in the Floridan Aquifer Due to Increased Withdrawals, Duval County, Florida", USGS/WRD/WRI-79/069, July 1979, U.S. Geological Survey, Tallahassee, Florida.

General Crude Oil Co. vs. Aiken (Groundwater Pollution by Saline Water from Oil Drilling Operations as Creating Permanent Damages), 335 S.W. 2D 229-235 (Civ. App. Tex. 1960).

Geraghty, J.J., et al., "Water Atlas of the United States", 1973, Water Information Center, Port Washington, New York.

Gill, H.E., "Ground Water Resources of Cape May County, N.J.,-Saltwater Invasion of Principal Aquifers", Special Report No. 18, 1962, U.S. Geological Survey and New Jersey Division of Water Policy and Supply, Trenton, New Jersey.

Grafton, R., "Salt-Water Intrusion in Southeastern Florida", Bulletin 3, October, 1968, Louisiana Water Resources Research Institute, Louisiana State University, Baton Rouge, Louisiana, pp. 15-30.

Gregg, D.O., "Protective Pumping to Reduce Aquifer Pollution, Glynn County, Georgia", Ground Water, Vol. 9, No. 5, September-October, 1971, pp. 21-29.

Groot, J.J., "Salinity Distribution and Ground-Water Circulation Beneath the Coastal Plain of Delaware and the Adjacent Continental Shelf", Open File Report No. 26, May, 1983, Delaware Geological Survey, Newark, Delaware.

Haire, W.J., "Hydrology of Western Collier County, Florida", 1979, U.S. Geological Survey, Miami, Florida.

Hampton, E.R., "Geology and Ground Water of the Molalla-Salem Slope Area, Northern Williamette Valley, Oregon", U.S. Geological Survey, Water-Supply Paper 1997, 1972.

Hardee, H.K., and McFadden, S.K., "Inventory of Known Ground Water Contamination Cases and Generalized Delineation of Five Ground Water Recharge Areas in South Carolina", March, 1982, Ground Water Protection Division, Department of Health and Environmental Control, Columbia, South Carolina.

Helwig, J.G., "Salt Water Intrusion in Florida (Legal Approaches to Prevention and Control)", 1969, Eastern Water Law Center, University of Florida, Gainesville, Florida.

Hughes, J.L., "Saltwater-Barrier Line in Florida: Concepts, Considerations, and Site Examples", USGS/WRD/WRI-79/073, June, 1979, U.S. Geological Survey, Tallahassee, Florida.

Jackson, W.T., and Paterson, A.M., "The Sacramento-San Joaquin Delta and the Evolution and Implementation of Water Policy. An Historical Perspective", W77-10101, June, 1977, Office of Water Research and Technology, U.S. Department of the Interior, Washington, D.C.

Jeffers, E.L., et al., "Water Reclamation and Automated Water Quality Monitoring', EPA-600/2-82-043, March, 1982, U.S. Environmental Protection Agency, Municipal Environmental Research Laboratory, Cincinnati, Ohio.

Jenks, J.H. and Harrison, B.L., "Multi-feature Reclamation Project Accomplishes Multi-Objectives", Water and Wastes Engineering, Vol. 14, No. 11, November, 1977, pp. 19-24, 30.

Johnson, M.J., "Ground-Water Conditions in the Eureka Area, Humboldt County, California, 1975", USGS/WRD/WRI-79/038, December, 1978, U.S. Geological Survey, Menlo Park, California.

Jorgensen, D.G., "Geohydrologic Models of the Houston District, Texas", Ground Water, Vol. 19, No. 4, July-August, 1981, pp. 418-428.

Kammerer, P.A., "Ground Water Quality Atlas of Wisconsin", Information Circular 39, February, 1981, U.S. Geological Survey, Madison, Wisconsin.

Kimmel, G.C., "Contamination of Ground Water by Sea Water Intrusion Along Puget Sound, Washington, An Area Having Abundant Precipitation", U.S. Geological Survey, Professional Paper 475-B, 1963.

Klein, H., et al., "Water and the South Florida Environment", Water-Resources Investigation 24-75, August, 1975, U.S. Geological Survey, Washington, D.C.

Kritz, G.J., "Analog Modeling to Determine the Fresh Water Availability on the Outer Banks of North Carolina", Report No. 64, April, 1972, North Carolina Water Resources Research Institute, North Carolina State University, Raleigh, North Carolina.

Land, L.F., "Ground-Water Resources of the Riviera Beach Area, Palm Beach County, Florida, USGS/WRD/WRI-77/076, September, 1977, U.S. Geological Survey, Tallahassee, Florida.

Land, L.F., "Effect of Lowering Interior Canal Stages on Salt-Water Intrusion into the Shallow Aquifer in Southeast Palm Beach County, Florida", Open File Report FL 75-74, 1975, U.S. Geological Survey, Tallahassee, Florida.

Larson, J.D., "Distribution of Saltwater in the Coastal Plain Aquifers of Virginia", Open File Report 81-1013, 1981, U.S. Geological Survey, Richmond, Virginia.

Lee, C., and Cheng, R.T., "On Seawater Encroachment in Coastal Aquifers", Water Resources Research, Vol. 10, No. 5, October, 1974, pp. 1039-1043.

Maloney, F.E., "Laws of Florida Governing Water Use", Journal of American Water Works Association, Vol. 47, No. 5, May, 1955, pp. 440-446.

Martin, P., "Ground-Water Monitoring at Santa Barbara, California, Phase 2 -- Effects of Pumping on Water Levels and Water Quality in the Santa Barbara Ground-Water Basin", Open File Report 82-366, April, 1982, U.S. Geological Survey, Sacramento, California.

Mattox, R.B., Saline Water, Proceedings of the Symposium on Groundwater Salinity, 46th Annual Meeting of the Southwestern and Rocky Mountain Division of the American Association for the Advancement of Science, Contribution No. 13 of the Committee on Desert and Arid Zones Research, Las Vegas, Nevada, 1970.

McCoy, J., "Hydrology of Western Collier County, Florida", Open File Report 72018, 1972, U.S. Geological Survey, Tallahassee, Florida.

McCoy, H.J., and Hardee, J., "Groundwater Resources of the Lower Hillsboro Canal Area, Southeastern Florida", Report of Investigation No. 55, 1970, Florida Bureau of Geology, U.S. Geological Survey, Tallahassee, Florida.

McDonald and Grefe, Inc., "Institutional and Financial Alternatives and Recommendations: AMBAG Section 208 Water Quality Management Plan", May, 1978, San Francisco, California.

McElwee, C.D., et al., "Salinity Control on the Smoky Hill River in Central Kansas", Salinity in Watercourses and Reservoirs, Proceedings of the 1983 International Symposium on State-of-the-Art Control of Salinity, July 13-15, Salt Lake City, Utah, Ed., French, R.H., Butterworth Publishers, Boston, Massachusetts, 1984, pp. 397-406.

Merritt, M.L., et al., "Subsurface Storage of Freshwater in South Florida: A Prospectus", Water Resources Investigation 83-4214, U.S. Geological Survey, 1983, Tallahassee, Florida.

Meyer, M., "A Brief Review of Ground Water Pollution Sources in South Dakota", South Dakota Department of Water and Natural Resources, Pierre, South Dakota, June, 1981.

Meyer, M., and Barwell, L., "Evaluation of Groundwater Resources, Eastern South Dakota and Upper Big Sioux River, South Dakota and Iowa: Task 5 - Water Quality Suitability by Aquifer for Drinking, Irrigation, Livestock Watering and Industrial Use", Vols. 1 and 2, South Dakota Department of Water and Natural Resources, Pierre, South Dakota, September, 1983.

Miller, D.W., DeLuca, F., and Tessier, T.L., "Ground Water Contamination in the Northeast States", EPA 660/12-74-056, June, 1974, U.S. Environmental Protection Agency, Washington, D.C.

Miller, D.W., Ed., "Waste Disposal Effects on Ground Water: A Comprehensive Survey of the Occurrence and Control of Ground Water Contamination Resulting from Waste Disposal Practices", Premier Press, Berkeley, California, 1980.

Miller, J.C., et al., "Groundwater Pollution Problems in the Southwestern United States:, EPA-600/3-77-012, January, 1977, Geraghty and Miller, Inc., Port Washington, New York.

Miller, J.C., "Ground-water Geology of the Delaware Atlantic Seashore", Report of Investigation No. 17, August, 1971, Delaware Geological Survey, Newark, Delaware.

Montanari, F.W., and Waterman, W.G., "Protecting Long Island Aquifers Against Salt-Water Intrusion", Bulletin 3, October, 1968, Louisiana Water Resources

Research Institute, Louisiana State University, Baton Rouge, Louisiana, pp. 59-86.

Muir, K.S., "Groundwater in the Seaside Area, Monterey County, California", Water Resources Investigations 82-10, September, 1982, U.S. Geological Survey, Sacramento, California.

Muir, K.S., "Seawater Intrusion and Potential Yield of Aquifers in the Soquel-Aptos Area, Santa Cruz County, California", AUSGS/WRD/WRI-81/020, October, 1980, U.S. Geological Survey, Menlo Park, California.

Muir, K.S., "Seawater Intrusion, Ground-Water Pumpage, Ground-Water Yield and Artificial Recharge of the Pajaro Valley Area, Santa Cruz and Monterey Counties, California", USGS/WRD-75-009, October, 1974, U.S. Geological Survey, Menlo Park, California.

Newport, B.D., "Salt Water Intrusion in the United States", EPA-600/8-77-011, July, 1977, Robert S. Kerr Environmental Research Laboratory, U.S. Environmental Protection Agency, Ada, Oklahoma.

Norris, S.E., "The Silurian Salt Deposits in Eastern Lake, Northwestern Ashtabula, and Northeastern Geauga Counties, Ohio", U.S. Geological Survey Open-File Report 79-269, December, 1978, Columbus, Ohio.

Oklahoma Water Resources Board, "Salt Water Detection in the Cimarron Terrace, Oklahoma", EPA-660/3-74-033, April, 1975, U.S. Environmental Protection Agency, Corvallis, Oregon.

Orange County Water District, "Annual Report: 50th Anniversary Edition", June, 1983, Orange County, California.

Owen, L.W., "The Challenge of Water Management: Orange County Water District, California", Bulletin 3, October, 1968, Louisiana Water Resources Research Institute, Louisiana State University, Baton Rouge, Louisiana, pp. 105-125.

Park, F.D.R., "Combating Saltwater Encroachment into the Biscayne Aquifer of Miami, Florida", Bulletin 3, October, 1968, Louisiana Water Resources Research Institute, Louisiana State University, Baton Rouge, Louisiana, pp. 31-56.

Peters, J.A., and Rose, J.L., "Water Conservation by Reclamation and Recharge", Journal of the Sanitary Engineering Division, American Society of Civil Engineers, Vol. 94, No. SA4, August, 1968, pp. 625-638.

Pickens vs. Harrison (Damages to Irrigation Well Caused by Salt Water Pollution from Oil Well), 246 S.W. 2D 316-319 (Civ. App. Tex. 1952).

Pinder, G.F., "Numerical Simulation of Salt-Water Intrusion in Coastal Aquifers, Part 1", OWRT-C-5224(4214)(2), 1975, Office of Water Research and Technology, U.S. Department of the Interior, Washington, D.C.

Price, D., Hart, D.H., and Foxworthy, B.L., "Artificial Recharge in Oregon and Washington", U.S. Geological Survey, Water-Supply Paper 1594-C, 1962.

Rodis, H.G., and Land, L.F., "The Shallow Aquifer -- A Prime Freshwater

Resource in Eastern Palm Beach County, Florida", USGS/WRD/WRI-76/038, February, 1976, U.S. Geological Survey, Tallahassee, Florida.

Rollo, J.R., "Salt Water Encroachment in Aquifers of the Baton Rouge Area-Louisiana", Water Resources Bulletin No. 13, August, 1969, Department of Conservation, Louisiana Geological Survey, Baton Rouge, Louisiana.

Rollo, J.R., "Ground Water in Louisiana", Water Resources Bulletin No. 1, August, 1960, Louisiana Geological Survey and Department of Public Works, Baton Rouge, Louisiana.

Rosenau, J.C., et al., "Geology and Ground-Water Resources of Salem County, New Jersey", Special Report No. 33, 1969, New Jersey Department of Conservation and Economic Development, Trenton, New Jersey.

Ryling, R.W., "A Preliminary Study of the Distribution of Saline Water in the Bedrock Aquifers of Eastern Wisconsin", Wisconsin Geological and Natural History Survey, University of Wisconsin, Madison, Wisconsin, 1961.

Schaefer, F.L., and Walker, R.L., "Saltwater Intrusion into the Old Bridge Aquifer in the Keyport-Union Beach Area of Monmouth County, New Jersey", Water Supply Paper 2184, 1981, U.S. Geological Survey, Washington, D.C.

Schiner, G.R., "Ground-Water Conditions in the Cocoa Well Field Area", 1980, U.S. Geological Survey, Orlando, Florida.

Schlicker, H.G., et al., "Environmental Geology of the Coastal Region of Tillamook and Clatsop Counties, Oregon", Oregon Department of Geology and Mineral Industries, Bulletin 74, 1970.

Science and Public Policy Program, "Ground Water Management in the Southeastern United States", EPA-600/2-83-090, September, 1983, U.S. Environmental Protection Agency, Washington, D.C.

Segol, G., and Pinder, G.F., "Transient Simulation of Saltwater Intrusion in Southeastern Florida", Water Resources Research, Vol. 12, No. 1, February 1976, pp. 65-70.

Shafer, G.H., "Groundwater Resources of Aransas County, Texas", Report 124, December, 1970, Texas Water Development Board, Austin, Texas.

Shafer, G.H., "Ground-Water Resources of Nueces and San Patricio Counties, Texas", Report 73, May, 1968, Texas Water Development Board, Austin, Texas.

Shedlock, R.J., "Saline Water at the Base of the Glacial-Outwash Aquifer Near Vincennes, Knox County, Indiana", USGS/WRD/WRI-80/067, August, 1980, U.S. Geological Survey, Indianapolis, Indiana.

Sherwood, C.B., and Klein, H., "Surface- and Ground-Water Relation in a Highly Permeable Environment", Extract of Publication No. 63, International Association of Scientific Hydrology, U.S. Geological Survey, Berkeley, California, 1963, pp. 454-468.

Sherwood, C.B., McCoy, H.J., and Galliher, C.F., "Water Resources of Broward

County, Florida", Report of Investigation No. 65, 1973, U.S. Geological Survey, Tallahassee, Florida.

Sloan, R.F., "Ground Water Resource Management in Kansas", Kansas Water News, Vol.22, No. 1 and 2, Winter, 1979, pp. 2-6, 23.

Smith, C.G., Jr., "Geohydrology of the Shallow Aquifers of Baton Rouge, Louisian", Bulletin GT-4, October, 1969, Louisiana Water Resources Research Institute, Louisiana State University, Baton Rouge, Louisiana.

Soren, J., "Hydrogeologic Conditions in the Town of Shelter Island, Suffolk County, Long Island, New York", USGS/WRD/WRI-78/024, February, 1978, U.S. Geological Survey, Syosset, New York.

Sproul, C.R., Boggess, D.H., and Woodard, H.J., "Saline Water Intrusion from Deep Artesian Sources in the McGregor Isles Area of Lee County, Florida", Information Circular No. 75, 1972, U. S. Geological Survey, Tallahassee, Florida.

Stegner, S.R., "Analog Model Study of Ground Water Flow in the Rehoboth Bay Area, Delaware", Technical Report No. 12, July, 1972, National Technical Information Service, U.S. Department of Commerce, Springfield, Virginia.

Steinkampf, W.C., "Origins and Distribution of Saline Ground Waters in the Floridan Aquifer in Coastal Southwest Florida", Water-Resources Investigations 82-4052, 1982, U. S. Geological Survey, Tallahassee, Florida.

Stewart, M.T., et al., "Evaluation of a Ground Transient Electromagnetic Remote Sensing Method for the Deep Detection and Monitoring of Salt Water Interfaces", Publication No. 73, 1982, Water Resources Research Center, University of South Florida, Tampa, Florida.

Swayze, L.J., "Hydrologic Base for Water Management, Dade County, Florida", 1979, U.S. Geological Survey, Miami, Florida.

Task Committee on Saltwater Intrusion, "Saltwater Intrusion in the United States", Journal of the Hydraulics Division, Proceeding of the American Society of Civil Engineers, Vol. 95, September, 1969.

Texas Water Development Board, "Continuing Water Resources Planning and Development for Texas, Phase I," Vol. I, May, 1977, Austin, Texas.

Todd, D.K., Ground Water Resources of the United States, Premier Press, Berkeley, California, 1983.

U.S. Geological Survey, National Water Summary 1983, 1984, Reston, Virginia.

U.S. Water Resources Council, "A Summary of Ground Water Problems", Report NTIS PB 82 160904, September, 1981, Washington, D.C.

Van der Leeden, F., Cerillo, L.A., and Miller, D.W., "Ground Water Pollution Problems in the Northwestern United States", EPA-660/3-75-018, U.S. Environmental Protection Agency, Corvallis, Oregon, May, 1975.

Wait, R.L, and Gregg, D.O., "Hydrology and Chloride Contamination of the

Principal Artesian Aquifer in Glynn County, Georgia", Hydrologic Report 1, 1973, Department of Natural Resources, Atlanta, Georgia.

Walter, G.R., Kidd, R.E., and Lamb, G.M., "Ground-Water Management Techniques for the Control of Salt-Water Encroachment in Gulf-Coast Aquifers", OWRT-B-073-ALA(1), 1979, Office of Water Research and Technology, U.S. Department of the Interior, Washington, D.C.

Walters, K.L., "Reconnaissance of Sea-Water Intrusion Along Coastal Washington, 1966-1968", Water Supply Bulletin No. 32, 1971, Washington Department of Ecology, Olympia, Washington.

Ward, F., "The Imperiled Everglades", National Geographic, Vol. 141, January, 1972, pp. 1-27.

Wasson, B.E., "Sources of Water Supplies in Mississippi", U.S. Geological Survey and Mississippi Research and Development Center, Jackson, Mississippi, 1980.

Wedding, J.J., Fong, M.A., and Crafts, C.M., "Use of Reclaimed Wastewater to Operate a Seawater Intrusion Control Barrier", Proceedings of Water Reuse Symposium, March 15-30, 1979, Washington, D.C., Vol. 1, 1979, American Water Works Association, Research Foundation, Denver, Colorado, pp. 639-662.

Whitfield, M.S., Jr., "Geohydrology and Water Quality of the Mississippi River Alluvial Aquifer, Northeastern Louisiana", Water Resources Technical Report No. 10, 1975, Louisiana Department of Public Works, Baton Rouge, Louisiana.

Wilmoth, B.M., "Development of Fresh Ground Water Near Salt Water in West Virginia", Ground Water, Vol. 13, No. 1, January-February, 1975, pp. 25-32.

Wilson, W.E., "Estimated Effects of Projected Ground-Water Withdrawals on Movement of the Saltwater Front in the Floridan Aquifer, 1976-2000, West-Central Florida", Water Supply Paper US 2189, 1982, U.S. Geological Survey, Washington, D.C.

Woodruff, K.D., "The Occurrence of Saline Ground Water in Delaware Aquifers", Report of Investigation No. 13, August, 1969, Delaware Geological Survey, Newark, Delaware.

CHAPTER 5

AREAS OF POTENTIAL PROBLEMS

This chapter combines the information presented in earlier chapters into a form which indicates potential salt-water intrusion problem areas in the contiguous United States. The areas are presented by the U.S. Water Resources Council's Water Resources Aggregated Subareas (ASA's). There are a total of 99 ASA's in the contiguous United States. Counties and states associated with each ASA are listed in Appendix B. This chapter describes the potential for salt-water intrusion in each ASA in terms of four categories: (1) salt-water intrusion is and will continue to be a problem; (2) salt-water intrusion is currently not a problem but there is a high potential for future problems; (3) salt-water intrusion is currently not a problem but there is a moderate potential for future problems; and (4) salt-water intrusion is currently not a problem, and there is only a low chance for future problems.

ASSUMPTIONS

The determination of the status of potential problem areas was made in a systematic manner by considering the information collected and summarized in Chapter 4. For each ASA the following information is depicted:

(1) Current Problem -- If an ASA is known to be experiencing salt-water intrusion problems. This information is obtained primarily from the "National Water Summary 1983" (U.S. Geological Survey, 1984). However, numerous other reports found in this study also indicated areas of salt-water intrusion, as did many of the telephone contacts. If any source of information indicated that a particular ASA was experiencing salt-water intrusion, that ASA is listed as experiencing a "Current Problem".

(2) Less Than 500 feet to Saline Water -- If the depth to saline ground water is less than 500 feet below the earth's surface, the potential for future salt-water intrusion problems increases. Furthermore, the greater the total dissolved solids (TDS) concentration is in the saline ground water, the greater is the sensitivity of the area's fresh ground water to salt-water intrusion (that is, less intrusion is required to contaminate the fresh ground water). Therefore, saline ground water is considered in two categories: (a) TDS is between 1,000 and 3,000 mg/l; and (b) TDS is greater than 3,000 mg/l. Each ASA is listed as to whether or not saline water is less than 500 feet below the earth's surface and in which of the two concentration categories the saline water falls. This information was obtained from a 1965 analysis of saline ground water in the United States (Feth et al., 1965).

(3) Future Problem -- If an ASA has been studied, and it has been recognized that future salt-water intrusion problems are likely to occur, that ASA is listed as such. This information comes from

numerous sources, including articles, books, technical reports, transactions or proceeding of conferences, and   the national telephone survey.

The "status" of each ASA was determined by considering the information available for each ASA. Since there are four types of information assembled for each ASA (current problems; less than 500 ft to saline ground water with TDS between 1,000 and 3,000 mg/l; less than 500 ft to saline ground water with TDS greater than 3,000 mg/l; and future problems) there are twelve possible combinations of the information. Any ASA which is currently experiencing salt-water intrusion, especially if the ASA has shallow saline ground water or future problems are recognized, is listed as having a "CONTINUING PROBLEM". If the ASA is not currently experiencing salt-water intrusion problems but future problems have been projected, it is listed as having a "HIGH POTENTIAL" for salt-water intrusion. If the ASA is not currently experiencing salt-water intrusion and future problems have not been projected, but it has shallow saline water, it is listed as having a "MODERATE POTENTIAL" for salt-water intrusion. Finally if there are no current or future salt-water intrusion problems, and there is not shallow saline ground water, the ASA is listed as having a "LOW POTENTIAL" for salt-water intrusion.

POTENTIAL PROBLEM AREAS

Table 10 contains a listing of the 99 ASA's in the contiguous United States and presents the status of salt-water intrusion of the nation. Figure 75 represents the data in a graphical format. Table 11 lists the total surface area and irrigated acres for each ASA. This information can be used to determine the potential magnitude of salt-water intrusion problems on agricultural activities.

Table 12 shows a summary of the number of telephone contacts and items of literature that cite specific problems in each state in the state-by-state survey. For example, in Alabama there were two references that cited sea-water intrusion as a problem, one that cited leaky wells, three that cited upconing and lateral intrusion, and two that cited natural salinity problems; in addition, there were two types of shallow saline ground water, and at least one reference predicted future problems. Two things should be noted about Table 12: (1) the number of listings in the state-by-state survey in Chapter 4 may not equal the number of citations given in Table 12 because some of the listings address more than one problem; and (2) the numbers in Table 12 represent the number of references found for each problem, not the number of different problem locations. References that deal with modeling or monitoring of a specific problem were included in the listing for that problem, while those that are less specific were included in the "Unspecified and Other Salinity Issue" column along with references for management, litigation, and general salt-water intrusion discussions.

The information from the Feth et al., (1965) analysis of saline ground water was the basis for the "Saline Water Less than 500 Ft Deep" column in Table 12. Water containing between 1,000 to 3,000 mg/l TDS and water containing more than 3000 mg/l TDS have again been separated for the same reasons that were cited for Table 10. As can be seen by the summation at the end of Table 12, 40 of the 48 contiguous states have shallow ground water containing between 1000 and 3000 mg/l TDS. In addition, 23 of these states

Table 10: Summary of Salt-water Intrusion Status by ASA

ASA	Current Problem	Less Than 500 Feet To Saline Water: TDS		Future Problem	Status
		1000-3000 mg/l	>3000 mg/l		
101					Low potential
102	X			X	Continuing problem
103	X			X	Continuing problem
104	X			X	Continuing problem
105	X			X	Continuing problem
106	X			X	Continuing problem
201		X		X	High potential
202	X	X		X	Continuing problem
203	X	X		X	Continuing problem
204		X			Moderate potential
205	X	X		X	Continuing problem
206				X	High potential
301	X	X		X	Continuing problem
302	X	X		X	Continuing problem
303	X			X	Continuing problem
304	X	X		X	Continuing problem
305	X	X		X	Continuing problem
306	X				Continuing problem
307	X	X		X	Continuing problem
308	X	X	X		Continuing problem
309	X			X	Continuing problem
401	X	X	X	X	Continuing problem
402	X	X	X		Continuing problem

Table 10: (Continued)

| ASA | Current Problem | Less Than 500 Feet To Saline Water: TDS | | Future Problem | Status |
		1000-3000 mg/1	>3000 mg/1		
403		X	X	X	High potential
404		X			Moderate potential
405		X	X		Moderate potential
406		X	X		Moderate potential
407	X	X		X	Continuing problem
408		X			Moderate potential
501		X		X	High potential
502		X		X	High potential
503		X			Moderate potential
504		X		X	High potential
505		X			Moderate potential
506		X		X	High potential
507		X	X		Moderate potential
601		X			Moderate potential
602		X	X		Moderate potential
701	X	X	X	X	Continuing problem
702					Low potential
703		X	X		Moderate potential
704	X	X	X		Continuing problem
705		X			Moderate potential
801	X	X			Continuing problem
802	X	X		X	Continuing problem
803	X	X		X	Continuing problem

Table 10: (Continued)

ASA	Current Problem	Less Than 500 Feet To Saline Water: TDS		Future Problem	Status
		1000–3000 mg/l	>3000 mg/l		
901	X	X	X	X	Continuing problem
1001	X	X	X	X	Continuing problem
1002	X	X	X	X	Continuing problem
1003	X	X	X	X	Continuing problem
1004	X	X	X	X	Continuing problem
1005	X	X	X	X	Continuing problem
1006	X	X		X	Continuing problem
1007		X			Moderate potential
1008		X			Moderate potential
1009		X	X		Moderate potential
1010	X	X	X		Continuing problem
1011	X	X	X		Continuing problem
1101	X	X			Continuing problem
1102	X	X	X		Continuing problem
1103	X	X	X		Continuing problem
1104	X	X			Continuing problem
1105		X	X		Moderate potential
1106	X		X	X	Continuing problem
1107		X			Moderate potential
1201	X	X			Continuing problem
1202	X	X		X	Continuing problem
1203	X	X	X		Continuing problem
1204	X	X	X		Continuing problem

Table 10: (Continued)

ASA	Current Problem	Less Than 500 Feet To Saline Water: TDS		Future Problem	Status
		1000-3000 mg/1	>3000 mg/1		
1205	X	X	X	X	Continuing problem
1301		X			Moderate potential
1302		X	X		Moderate potential
1303		X	X		Moderate potential
1304	X	X	X		Continuing problem
1305		X	X		Moderate potential
1401	X	X	X		Continuing problem
1402		X	X		Moderate potential
1403		X	X		Moderate potential
1501	X	X	X		Continuing problem
1502	X	X	X		Continuing problem
1503	X	X	X		Continuing problem
1601		X	X	X	High potential
1602		X	X	X	High potential
1603		X	X		Moderate potential
1604		X	X		Moderate potential
1701					Low potential
1702	X	X	X		Continuing problem
1703		X	X		Moderate potential
1704					Low potential
1705	X	X	X		Continuing problem
1706	X	X	X	X	Continuing problem
1707		X	X		Moderate potential

Table 10: (Continued)

ASA	Current Problem	Less Than 500 Feet To Saline Water: TDS		Future Problem	Status
		1000-3000 mg/l	>3000 mg/l		
1801					Low potential
1802					Low potential
1803		X	X		Moderate potential
1804	X	X			Continuing problem
1805	X	X	X	X	Continuing problem
1806	X	X	X		Continuing problem
1807		X	X		Moderate potential

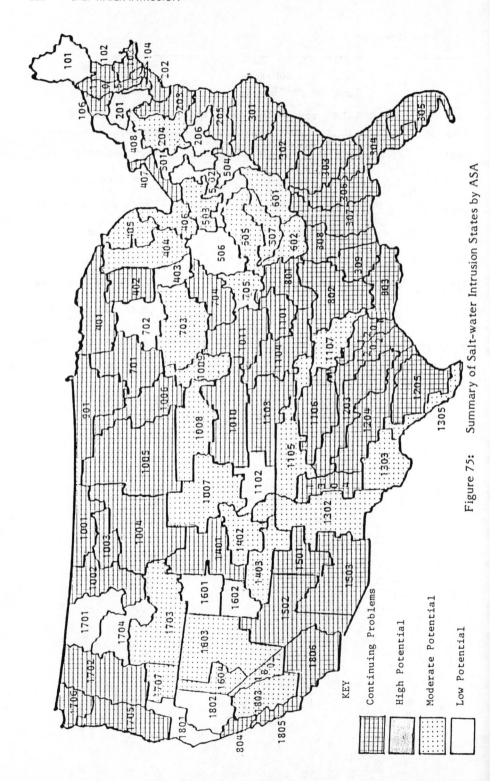

Figure 75:   Summary of Salt-water Intrusion States by ASA

KEY

Continuing Problems

High Potential

Moderate Potential

Low Potential

Table 11: Total Surface Area and Irrigated Acres
           for Each ASA

ASA	Total Surface Area	Total Irrigated Lands
	(1000 Acres)	
101	19968.4	3.0
102	4147.9	11.1
103	4253.1	21.9
104	3088.1	0.7
105	7936.3	9.1
106	4783.5	1.4
201	9712.5	4.1
202	3114.7	40.3
203	9908.7	127.4
204	17236.5	18.2
205	16240.6	100.3
206	9634.1	18.9
301	25668.9	306.2
302	27126.9	96.2
303	22800.8	303.9
304	18788.3	1060.7
305	12535.9	1502.2
306	18325.1	619.5
307	20889.8	75.9
308	15349.4	11.7
309	12840.0	0.0
401	17152.4	5.0
402	11227.7	119.2
403	5304.6	59.1
404	16845.5	389.9
405	8665.2	36.2
406	10468.7	25.1
407	5415.8	5.7
408	10952.4	13.5
501	9759.9	0.0
502	17932.5	4.8
503	13062.3	18.4
504	8312.9	0.0
505	20414.4	19.8
506	21538.2	65.1
507	11377.1	5.0
601	15764.5	10.8
602	11491.7	8.8

Table 11: (Continued)

ASA	Total Surface Area	Total Irrigated Lands
	(1000 Acres)	
701	28683.4	354.2
702	19735.9	185.2
703	35630.6	111.1
704	19601.6	138.0
705	12056.2	15.5
801	17959.6	2855.6
802	31195.2	1422.5
803	17252.9	1052.7
901	34987.9	94.7
1001	16406.8	205.8
1002	24043.2	998.3
1003	11017.7	193.3
1004	45828.6	1286.8
1005	61064.5	399.4
1006	23914.8	389.6
1007	38238.9	2079.2
1008	29950.1	3892.2
1009	14269.8	200.1
1010	35291.4	4003.4
1011	26261.8	120.2
1101	12810.9	281.6
1102	15841.7	519.2
1103	29680.1	2836.8
1104	25504.3	138.6
1105	29951.8	2051.0
1106	25824.5	1367.2
1107	16483.8	80.6
1201	11273.5	155.4
1202	17037.9	629.2
1203	29795.3	3535.8
1204	29988.5	1628.7
1205	25363.0	654.4
1301	5254.4	618.7
1302	34637.0	669.2
1303	21175.4	280.5
1304	13039.9	253.1
1305	10551.1	755.0

Table 11: (Continued)

ASA	Total Surface Area	Total Irrigated Lands
	(1000 Acres)	
1401	29675.6	856.7
1402	16523.9	502.6
1403	19671.9	313.2
1501	17042.8	113.6
1502	40523.4	162.1
1053	44506.9	1154.9
1601	17369.8	902.4
1602	13069.3	436.1
1603	46896.2	805.4
1604	11878.3	264.7
1701	23183.8	708.1
1702	38458.1	2250.2
1703	44014.2	4100.5
1704	20249.7	307.6
1705	24921.9	524.6
1706	10565.6	74.4
1707	11896.1	492.1
1801	15272.5	481.7
1802	20692.3	2297.5
1803	21022.4	5608.0
1804	4807.6	301.8
1805	7201.3	471.5
1806	27473.6	806.3
1807	9028.7	62.2

Table 12: Delineation of Specific State Salt-water Intrusion Problems

State	Sea Water Intrusion	Leaking Wells	Up-Coning	Natural Salinity	Unspecified & Other Salinity Issue	Saline Water less than 500 ft deep 1000-3000 mg/l (TDS)	>3000 mg/l (TDS)	Future Problems Cited
Alabama	2*	1	3	2		X	X	yes
Arizona				1		X	X	
Arkansas		1	2			X		yes
California	17			2	5	X	X	yes
Colorado						X	X	yes
Connecticut	5							yes
Delaware	7		3			X		
Florida	17	4	5	4	14	X		yes
Georgia	7			1	2	X		yes
Idaho				1	3			
Illinois			3	2	1	X		yes
Indiana		1	1	1	1	X		yes
Iowa		1	1	4		X	X	yes
Kansas		1	2	4	1	X	X	

Table 12: (Continued)

State	Sea Water Intrusion	Leaking Wells	Up-Coning	Natural Salinity	Unspecified & Other Salinity Issue	Saline Water less than 500 ft deep 1000-3000 mg/1 (TDS)	>3000 mg/1 (TDS)	Future Problems Cited
Kentucky		2				X		
Louisiana	1			3	3	X		yes
Maine	2		3					
Maryland	4	2			2			yes
Massachusetts	5				1			yes
Michigan		2	1	1		X	X	yes
Minnesota			1	2		X	X	yes
Mississippi	1	3		4		X		yes
Missouri				3		X	X	yes
Montana		2		1	2	X	X	yes
Nebraska				2		X	X	yes
Nevada						X	X	
New Hampshire	4							yes
New Jersey	7	1	8	1	2	X		yes

Table 12: (Continued)

State	Sea Water Intrusion	Leaking Wells	Up-Coning	Natural Salinity	Unspecified & Other Salinity Issue	Saline Water less than 500 ft deep 1000-3000 mg/1 (TDS)	>3000 mg/1 (TDS)	Future Problems Cited
New Mexico		1	3	1		X	X	yes
New York	13	2	2		2	X		yes
North Carolina	4	1	1	1	1	X		yes
North Dakota		1	1	3		X	X	yes
Ohio		3		1		X		yes
Oklahoma		5		2		X	X	yes
Oregon	4		3		2	X	X	
Pennsylvania		2	5		3	X		yes
Rhode Island	5							
South Carolina	3		1	2	2	X		yes
South Dakota		1	1	4		X	X	yes
Tennessee						X	X	
Texas	10	2	1	3		X	X	yes
Utah				2		X	X	yes

Table 12: (Continued)

State	Sea Water Intrusion	Leaking Wells	Up-Coning	Natural Salinity	Unspecified & Other Salinity Issue	Saline Water less than 500 ft deep		Future Problems Cited
						1000-3000 mg/1 (TDS)	> 3000 mg/1 (TDS)	
Vermont				1	1			yes
Virginia	2		2		1	X		yes
Washington	4				3	X	X	yes
West Virginia					2	X		yes
Wisconsin			1	3	1	X	X	
Wyoming		2		1		X	X	yes
TOTAL NUMBER OF STATES	21	20	22	30	22	40	23	36

* numbers in table denote numbers of reference items

have water containing over 3000 mg/l TDS. It follows logically that natural salinity is the most frequently addressed problem. The other problems seem to be equally distributed throughout the nation, with the exception of New England, which is primarily experiencing sea-water intrusion in the coastal areas.

CONCLUSIONS

Several conclusions can be reached from these data. First, it is somewhat surprising that problems associated with irrigation and the increasing salinity in soil and surface water found in the arid western states were not more correlated to ground water salinity problems. This may be due to the fact that these arid areas are importing the irrigation water from areas other than where the crops are grown. The indication is that there is little or no ground water in the crop-land areas. Therefore, any increasing salinity in what little ground water there may be is not seen as a problem and subsequently not documented.

Second, sea-water intrusion has occurred in each of the 21 coastal states. Although the problem is primarily associated with population size (water demand) rather than agricultural activities, it is clear that sea-water intrusion is a substantial threat when ground water basins in hydrological connection with the sea are developed. As water demand increases and water supply decreases, agricultural activities will be affected.

The noncoastal states (as well as inland areas of coastal states) also face ground water contamination from salt water. Brine disposal, leaking wells, or abandoned wells are often cited as the major problem in many areas. Although these problems appear to be site specific and localized, mineral exploration and production are often conducted in a regional manner. These regional activities can and do affect the quality of ground water, reducing the usefulness of the associated aquifers.

Finally, only seven states do not overlay shallow saline water. These few states can develop their aquifers without fear of upconing, but they must still recognize the potential threat of brine disposal or other salt related activities. The other 41 states should remain cautious to the fact that overpumping ground water may ultimately result in an unusable aquifer due to intrusion from shallow saline ground water.

In summary, salt-water intrusion can be seen as a wide-spread problem. Since all projections indicate an increase in ground water use, more salt-water intrusion can be expected unless proper and careful planning of ground water utilization precedes development. Although all salt-water intrusion incidents are not caused by agricultural activities, agriculture will be impacted by all salt-water intrusion. The keys to controlling ground water deterioration from salt-water intrusion are to view ground water as a national resource which crosses state boundaries and to plan for its multiple usage for meeting domestic, agricultural, municipal, and industrial needs.

SELECTED REFERENCES

Feth, J.H. et al., "Preliminary Map of the Conterminous U.S. Showing Depth to and Quality of Shallowest Ground Water Containing More Than 1000 Parts Per

Million Dissolved Solids", Atlas HA-199, 1965, U.S.   Geological Survey, Washington, D.C.

U.S. Geological Survey, National Water Summary 1983, 1984, Reston, Virginia.

APPENDIX A

ANNOTATED BIBLIOGRAPHY ON SALT WATER INTRUSION

This annotated bibliography is based on a computer-based search of the literature using key descriptor words. A total of ten bibliographic bases were searched via the Dialog system: Agricola (1970-1978); Agricola (1979-1984); Compendex; Conference Papers Index; CRIS/USDA; Dissertation Abstracts; National Technical Information Service; Pollution Abstracts; Smithsonian Scientific Information Exchange Current Research; and Water Resources Abstracts. A total of 672 abstracts were identified using key words such as salt water intrusion, salt water encroachment, and ground water. The total number of abstracts were screened for relevancy and to eliminate duplication. As a result, 125 abstracts are included in this appendix. Each included abstract is categorized into one of the following categories: bibliographic materials (five abstracts); overview information on salt water intrusion (thirteen abstracts); coastal salt water intrusion (twenty-seven abstracts); noncoastal intrusion of salt water (eight abstracts); oil-gas field activities (eight abstracts); irrigation effects (three abstracts); fresh water (injection and surface water pressure-head) as a barrier to seawater (twelve abstracts); institutional measures (eight abstracts); modeling: coastal seawater intrusion--physical (eight abstracts); modeling: coastal seawater intrusion--numerical (twenty-two abstracts); modeling: noncoastal salt water intrusion--physical (three abstracts); and modeling: noncoastal salt water intrusion--numerical (eight abstracts). It should be noted that a given abstract could possibly have been placed in more than one category; however, only one was used. Therefore, abstracts with information pertaining to a particular category could be found in other categories.

BIBLIOGRAPHIES

AB-1.   Lehmann, E.J., "Studies of Saline Soils and Ground Water. A Bibliography with Abstracts (1964-Oct. 1973)", NTIS-WIN-73-076, November, 1973, National Technical Information Service, U.S. Department of Commerce, Springfield, Virginia.

> This NTISearch bibliography contains 55 selected abstracts of research reports retrieved using the NTIS on-line search system--NTISearch. The abstracts cover the following topics on saline soils and ground water: irrigation return flow, hydrology, aquifers, water pollution, salt water intrusion, and effects upon water supplies.

AB-2.   Lehmann, E.J., "Ground Water Pollution. Part 3. Saline Ground Water (Citations from the NTIS Data Base) (1964-Jan. 1977)", NTIS/PS-77/0009/9, January, 1977, National Technical Information Service, U.S. Department of Commerce, Springfield, Virginia.

> The abstracts cover federally-funded research on topics dealing with saline ground water: irrigation return flow, soil studies, hydrology, aquifers, water quality, salt water intrusion, and effects upon water supplies. This updated bibliography contains 121 abstracts, nine of which are new entries to the previous edition.

AB-3.   National Technical Information Service, "Ground Water Pollution: Saline Ground Water. 1964-May, 1981 (Citations from the NTIS Data

Base)", PB83-800813, November, 1982, U.S. Department of Commerce, Springfield, Virginia.

The abstracts cover federally-funded research on topics dealing with saline ground water: irrigation return flow, soil studies, hydrology, aquifers, water quality, salt water intrusion, and effects upon water supplies. This updated bibliography contains 261 citations, none of which are new entries to the previous edition.

AB-4.   National Technical Information Service, "Ground Water Pollution: Saline Ground Water. June, 1981-September 1982 (Citations from the NTIS Data Base)", PB83-800821, November, 1982, U.S. Department of Commerce, Springfield, Virginia.

The abstracts cover federally-funded research on topics dealing with saline ground water: irrigation return flow, soil studies, hydrology, aquifers, water quality, salt water intrusion, and effects upon water supplies. This updated bibliography contains 39 citations, all of which are new entries to the previous edition.

AB-5.   Resources Scientific Information Center, "Subsurface Water Pollution. A Selective Annotated Bibliography. Part II. Saline Water Intrusion", WRSIC-72-221E, March, 1972, Office of Water Resources Research, U.S. Department of the Interior, Washington, D.C.

Subsurface water pollution is a selective bibliography produced by the U.S. Environmental Protection Agency from the computerized data base of the Water Resources Scientific Information Center, U.S. Department of the Interior. This bibliography represents published research and development in water resources as abstracted and indexed in the semimonthly journal Selected Water Resources Abstracts (SWRA). This bibliography represents a search of a 33,980-item data base, covering SWRA from October, 1968, through December, 1971, and is published in three parts. Part II covers pollution associated with saline water intrusion. The bibliography contains references to technology dealing with the prevention or abatement of pollution, litigation pertaining to incidences of pollution, and laws and regulations pertaining to the construction and operation of subsurface waste disposal facilities.

OVERVIEW INFORMATION ON SALT WATER INTRUSION

AB-6.   Anonymous, "Sea-Water Intrusion in California. Inventory of Coastal Ground Water Basins", Bulletin N-63-5, October, 1975, California Department of Water Resources, Sacramento, California.

This bulletin contains a series of reports and discusses the physical characteristics of California's coastal ground water basins, including geology and ground water hydrology and water quality, status of sea-water intrusion, methods of control, experimental studies, and preliminary plans for prevention and control of intrusion.

AB-7.   Bruington, A.E., "Saltwater Intrusion into Aquifers", Water Resources Bulletin, Vol. 8, No. 1, February, 1972, pp. 150-160.

Salt-water intrusion into fresh ground water aquifers is common throughout the United States. There are three mechanisms by which intrusion occurs. One is the reversal or reduction of ground water gradients, which allows saline water to move. A second is through the destruction of natural barriers that formerly prevented the movement of salt waters or separated bodies of fresh and salt water. The third is the improper disposal of waste saline water. One solution to the problem is the pressure barrier system used in Los Angeles County, where a line of injection wells was installed parallel to the seacoast. Fresh water is injected into these wells under pressure so that a pressure barrier is formed high enough to control intruding sea water. The pumping barrier is a line of pumping wells between the source of intrusion and the inland pumping fields. These wells are pumped at a rate which will cause a ground water pressure trough to be formed along the line of wells low enough so that sea water cannot pass inland. Since the problem usually stems from over-pumping, a program of reduced pumping can be developed. Another way is to continue pumping in the historical pattern but increase the available ground water supply by recharge.

AB-8.   Eldridge, E.F., "Report of Research and Research Needs on Water Quality Problems in the State of Hawaii", June, 1963, U.S. Public Health Service, Washington, D.C.

A research program designed to provide the necessary knowledge to meet existing and potential problems of water quality in Hawaii is presented. Existing water quality problems include: (1) contamination of subsurface sources of domestic water supply, especially basal, stemming from salt water intrusion, effluent bacteria and virus, fertilizers, pesticides and herbicides, and land disposal of refuse; (2) pollution of nearshore ocean waters caused by ocean disposal of effluent from sewage and waste treatment facilities. Industrial pollution sources are sugar mills, pineapple canneries, and an oil refinery. Recommended research studies are outlined as follows: (1) studies of basal water problems--salt water intrusion, ground water movement, fate and travel of bacteria and virus, fate and travel of fertilizers, insecticides and herbicides, and reactivity of soils and heavy metals; (2) ocean disposal of effluents--oceanographic study, and effluent-salt water mixing; (3) water quality versus fish and shellfish--field investigation, bioassay and shellfish contamination; (4) sewerage--sewage disposal by irrigation, sewage lagoons, combined sewage and industrial wastes, and sewerage planning; (5) industrial waste disposal--cane sugar mills and pineapple cannery wastes; (6) salt water conversion; and (7) field and laboratory methods--organics, heavy metals and soils.

AB-9.   Fetter, C.W., Jr., "Water Resources Management in Coastal Plain Aquifers", Proceedings of First World Congress on Water Resources, Sept. 24-28, 1973, Chicago, Illinois, Vol. 1, International Water Resources Association, Champaign, Illinois, 1973, pp. 322-331.

Water quality degradation, saline water encroachment and disruption of coastal zone ecosystems have occured in a number of localities. All of these problems can be overcome through proper water resources management.

AB-10. Florida House of Representatives, "Report of the Select Committee on Water Resources (Foreign Matter and Salt Water Intrusion in Public Waters)", Tallahassee, Florida, October, 1964, pp. 12-15.

Navigation and recreational uses of public waters have suffered from deposits of foreign matter in the waters. Sunken vessels and other debris are unsightly and dangerous to sportsmen and recreationists. The committee recommends that legislation be passed prohibiting such deposit of any foreign matter into public waters. A further problem affecting water supplies in coastal areas is the intrusion of salt water. The primary factors contributing to this intrusion are: (1) loss of fresh water pressure through increased demands of municipalities, agriculture, and industry; (2) excessive drainage; (3) lack of protective works against tidewaters in bays, canals, and rivers; (4) improper location of water wells; and (5) highly variable annual rainfall with insufficient surface storage during droughts. One method of reducing this intrusion requires the establishing of salt barrier lines along coastal areas. No canals or streams could be constructed or enlarged inland from this line unless structures are installed to prevent salt water from moving inland. The committee recommends that the board of conservation be required to establish these lines and that permits be required before any canal can be constructed or any stream enlarged which would discharge into tidal waters.

AB-11. Florida House of Representatives, "Report of the Select Committee on Water Resources (Florida's Water Problems)", Tallahassee, Florida, October, 1964.

The Select Committee on Water Resources conducted meetings at various locations around the state in order to familiarize its members with Florida's many water problems. The problems under consideration included: (1) industrial pollution; (2) domestic pollution; (3) deposit of foreign matter in public waters; (4) salt water intrusion; (5) identification and designation of floodplain areas; (6) financing acquisition of water storage lands; (7) licensing of well drillers; (8) hard detergents; and (9) codification of drainage laws. The interest generated at the hearings centered around the general topics of pollution abatement and enforcement and implementation of existing laws. The committee made various recommendations in each of these areas and proposed legislation to carry out the recommendations. The committee also suggested that further research be authorized either by a special study group or by an interim committee to report to the next legislature.

AB-12. Kashef, A.I., "What Do We Know About Salt Water Intrusion?", Water Resources Bulletin, Vol. 8, No. 2, April, 1972, pp. 282-293.

A literature review of the main investigations in the field of salt water intrusion problems is summarized. The discussions are subdivided into the period through which the problems were identified and field observations were explained; the analytic approaches during the period from 1940 to the late sixties; and the refined techniques during the past three years. A summary of the recharge methods is also given.

AB-13. LeGrand, H.E., "Role of Ground Water Contamination in Water Management", Journal of the American Water Works Association, Vol. 59, No. 5, May, 1967, pp. 557-565.

Ground water contamination and its complement are placed in perspective with water development. Their importance in long-range integrated plans is demonstrated. Ground water contamination involves quality deterioration of underground supplies. These are influenced by waste disposal practices, artificial recharge of aquifers, accidents, and salt water intrusion. A major objective is to maintain good health practices and minimize disposal costs. Variable factors for pollution are the variety of waste materials, wide toxicity ranges, man's disposal patterns and accidental releases, subsurface water development, elemental behavior in soil, water, and rock environment, and spatial and temporal criteria. Balanced surface and ground water development and waste disposal policies are needed. Other considerations and land and stream classifications are outlined.

AB-14. Sloan, R.F., "Ground Water Resource Management in Kansas", Kansas Water News, Vol. 22, No. 1 and 2, Winter, 1979, pp. 2-6, 23.

The Kansas Ground Water Management District Act of 1972 encouraged the formation of local ground water management districts each with its own unique hydrogeologic characteristics. Local decision making would be supported by state guidance, technical expertise and financial assistance to obtain the optimum product of most benefits/least costs. A brief synopsis of five Kansas ground water management districts is given. Ground water quantity and quality problems have given rise to associated management strategies suitable for each district. This has resulted in stabilization of ground water levels or decelerating drawdown to acceptable rates, monitoring of salt water intrusion, irrigation scheduling for optimum ground water use, and augmenting available ground water supplies through weather modification and artificial recharge.

AB-15. U.S. Environmental Protection Agency, "Identification and Control of Pollution from Salt Water Intrusion", EPA/430/9-73-013, 1973, Office of Air and Water Programs, Washington, D.C.

Saltwater intrusion, whether into surface water or ground water, is a complex situation controlled by the geologic and hydrologic characteristics of the area. Natural water systems are dynamic. They respond in quality and quantity to natural phenomena and to man's activities such as changes in land use, stream channel linings, and consumptive withdrawal. Identification and evaluation of the nature and extent of saltwater intrusion begins with an understanding of the general mechanisms by which intrusion occurs. A basic framework is given for assessment of saltwater intrusion problems and their relationship to the total hydrologic system, to aid state authorities in developing areawide waste treatment management plans.

AB-16.   Ward, F., "The Imperiled Everglades", National Geographic, Vol. 141, January, 1972, pp. 1-27.

The series of rampaging fires which destroyed large parts of the Everglades in 1971 were the result of a larger crisis which threatens the Everglades and the entire population of South Florida. The crisis is water--how it is managed and for what purpose. The Everglades suffered through a severe drought in 1970 and 1971. The activities of man in draining and developing the area magnified the adverse effects of the drought. The entire Everglades area was once covered by water. The flow of water over the Everglades is essential to the prevention of salt water intrusion into the aquifer from which the Miami area draws its drinking water. However, water levels have been upset both above and below the surface. Development in the Big Cypress swamp from which the Everglades receives 56 percent of its surface waters is partially to blame. But the major problem stems from the drainage of all the land south and east of Lake Okeechobee except for the Everglades Park and certain water conservation areas, for the purpose of flood control. Proposed solutions are reflooding the entire area as well as raising the level of Lake Okeechobee.

AB-17.   Wasserman, L.P., "Sweetwater Pollution", Science and Technology, No. 90, June, 1969, pp. 20-27.

The earth's fresh water supply is not a constant and is being reduced by man's wanton use. The problem arises from pollution and wasted water and from insidious circumstances resulting from such activities as land filling along shorelines and marshes. Pollution problems from sewage, industrial wastes, agricultural chemicals, and eutrophication are reviewed. Landfills of marshes, estuaries, or river shores produce adverse ecological changes, destroying valuable breeding grounds for many forms of aquatic life, waterfowl, and other wildlife. Radioactive materials released in even minute quantities present a great hazard because the radioactive material can become progressively concentrated to dangerous levels in various forms of animal life. Thermal pollution, sedimentation, and salt water intrusion problems are discussed. Current approaches to pollution problems are merely stopgap approaches. The principal ingredient needed to counteract the growing water pollution dilemma is money--money to run a conservation organization, to preserve and set aside lands, to run treatment plants, and to coordinate plans on national and state levels.

AB-18.   Young, R.H.F., "Pollution Effects on Surface and Ground Waters, (Literature Review)", Journal of the Water Pollution Control Federation, Vol. 46, No. 6, June, 1974, pp. 1419-1429.

Literature concerning the pollution effects of various substances on surface and ground water is reviewed. Among the substances covered are: nutrients, agricultural wastes, chemicals, heavy metals and radionuclides, and biological contamination. Nutrient enrichment sources cited were sewage treatment effluents, industrial wastes, urban runoff, and agricultural runoff. Documented sources of agricultural pollution were: (1) percolates from surface irrigated dairy manure slurries, (2) storm runoff from cattle feedlots, (3) runoff from agricultural watersheds, and (4) seepage from wastewater irrigation. Chemical pollution sources were: oil field brine disposal; salt-water intrusion in coastal areas; irrigation-return flow; contaminants from outboard motor fuel; herbicides; use of deicing salts on highways; and the mobilization of the constituents in contaminated snow, such as heavy metals, oils, greases, phenols, and BOD from decaying organic matter. Heavy metal and radionuclide contamination sources were discharges from gold recovery operations, use of nuclear reactors, and nuclear weapons tests. Sources of biological contamination included: (1) slime outbreaks due to industrial or domestic wastewater effluents, (2) coliforms due to discharges from boats and a faulty septic tank, and (3) viruses from septage filtrates. Reclamation by ground water recharge, soil pollution, and modeling analytical research methods were reviewed.

COASTAL SALTWATER INTRUSION

AB-19.   Anderson, M.P. and Berkebile, C.A., "Evidence of Salt-Water Intrusion in Southeastern Long Island", Ground Water, Vol. 14, No. 5, September-October, 1976, pp. 315-319.

A preexisting network of 124 wells used for fire protection in the Town of Southampton, Long Island, New York, was monitored during a one-year period for ground water levels and chloride concentrations. Water from 26 wells had chloride concentrations of 50 mg/l or greater, and in 17 of the wells, the chloride concentration exceeded 200 mg/l. Of the mentioned 26 wells, 19 were located in two densely populated areas. The possibility of widespread intrusion in the wells of one of the densely populated areas had not been previously documented. In addition, detailed water table contour maps were presented for selected areas within the town.

AB-20.   Buxton, H.T. et al., "Reconnaissance of the Ground-Water Resources of Kings and Queens Counties, New York", Open-File Report 81-1186,

1981, U.S. Geological Survey, Syosset, New York.

Western Long Island's ground water resources are being considered for redevelopment as a supplemental source of public water supply. The ground water reservoir in Kings and Queens Counties had supplied an average of more than 120 mgd for industrial and public water supply from 1904 to 1947. However, deterioration of ground water quality from induced encroachment of salty ground water caused the total cessation of pumping for public supply in Kings County in 1947 and in western Queens County in 1974. This investigation was designed to make a preliminary evaluation of the present ground water quality and to review hydrologic information pertinent to the formulation of an appropriate development plan. Detailed hydrogeologic mapping was done to define the geometry of the ground water system. The history of ground water development and the consequent responses of ground water levels and quantity were summarized. A well network was established to monitor ground water levels and ground water quality in Kings and Queens Counties. The distribution of chloride and nitrate in ground water was used to identify contamination from salt water intrusion and surface sources. A limited source of potable ground water is available in the Magothy-Jameco and Lloyd Aquifers; however, an accurate estimate of the quantity of water available or a draft rate suitable for maintaining a usable ground water quality is undetermined.

AB-21. Fairchild, R.W. and Bentley, C.B., "Saline-Water Intrusion in the Floridan Aquifer in the Fernandina Beach Area, Nassau County, Florida", USGS/WRD/WRI-77/075, September, 1977, U.S. Geological Survey, Tallahassee, Florida.

The Floridan Aquifer is the major source of water for municipal and industrial use in northeastern Nassau County, Florida. Pumpage from the aquifer at Fernandina Beach has created a cone of depression in the potentiometric surface which is at or below sea level in a 35-square mile area. Since 1880 water levels have declined more than 120 feet near the center of pumping, and chloride concentrations have increased from about 20 to a maximum of 800 mg/l in some deep wells. The saline water probably is connate water which comes from zones below 1,250 and 2,000 feet. Several alternate methods to control the migration of saline water are discussed.

AB-22. Grafton, R., "Salt-Water Intrusion in Southeastern Florida", Bulletin 3, October, 1968, Louisiana Water Resources Research Institute, Louisiana State University, Baton Rouge, Louisiana, pp. 15-30.

Overdrainage as a result of urbanization and geological conditions resulted in salt-water intrusion into southeastern Florida. After massive flood damage in 1957 and 1958, the central and southern Florida Flood Control District was established to acquire lands, represent local interests, raise funds, and operate projects in cooperation with the Federal government to promote flood control and water preservation.

Further legislation is required to broaden the scope of the F.C.D., especially in salt-water intrusion. With the exception of Dade and Broward Counties, little is being accomplished by other counties. Increased population is endangering lands committed to conservation areas. However, activities by special agencies against salt-water intrusion in southeastern Florida continue to gain momentum.

AB-23.   Haire, W.J., "Hydrology of Western Collier County, Florida", 1979, U.S. Geological Survey, Miami, Florida.

Continued rapid increases in population have imposed large demands for fresh water on the Naples Municipal Water System. Extension of the well field along the coastal ridge will meet immediate and near future needs but fresh ground water supplies to meet predicted needs in 1985 will have to come from inland sources. This inland source must be protected from over drainage by existing canals, contamination resulting from land development, and salt water intrusion. The results from this investigation will determine the locations which would most likely yield the greatest quantities of the best quality water to supply the ultimate municipal needs of western Collier County. Supplemental data will indicate the effect of the canal system on aquifer storage. These objectives will be obtained by a reconnaissance of the two-square mile area east of Naples to determine exact locations and extent of existing canals and control structures and all works proposed for the near future; a series of periodic streamflow measurements at several locations in canals to determine incremental losses or increases of water within specific reaches of the canals; exploratory drilling for geologic information and more detailed information on the permeability of the shallow aquifer; and water sampling of canals and wells to determine changes in water quality with time.

AB-24.   Hughes, J.L., "Saltwater-Barrier Line in Florida: Concepts, Consider-ations, and Site Examples", USGS/WRD/WRI-79/073, June, 1979, U.S. Geological Survey, Tallahassee, Florida.

Construction of canals and enlargement of streams in Florida have been mostly to alleviate the impact of floods and to drain wetlands for development. Land drainage and heavy pumpage from coastal water-table aquifers have degraded potable ground and surface water with saltwater. Control of saltwater intrusion is possible through implementation of certain hydrologic principles. This report briefly reviews the fundamentals of saltwater intrusion in surface-water channels and associated coastal aquifers, describes the effects of established saltwater-barrier lines in Florida, and gives a history of the use and benefits of salinity-control structures.

AB-25.   Johnson, M.J., "Ground-Water Conditions in the Eureka Area, Humboldt County, California, 1975", USGS/WRD/WRI-79/038, December, 1978, U.S. Geological Survey, Menlo Park, California.

Ground water conditions in the Eureka area were evaluated during 1975 to determine whether significant changes had occurred since 1952, when an earlier reconnaissance was made. No major changes in water levels or water quality were noted at 1975 pumping rates. Recharge to the ground water system compensates for both artificial and natural discharge. The position of the freshwater-saltwater transition zone underlying the alluvial flood plains in the summer of 1975 were unchanged from those in 1952. Ground water continues to be used principally for irrigation on the alluvial flood plains of the Eel and Mad Rivers. In 1975, about 425 irrigation wells supplied an estimated 24,000 acre-feet of ground water to the Eureka area. The estimated total ground water pumpage for irrigation, industry, public supply, domestic use, and livestock was 27,500 acre-feet. During the irrigation season, there is a higher rate of recharge from the rivers which helps to maintain ground water storage and stabilize the freshwater-seawater transition zone.

AB-26.   Kashef, A.I., "Salt Water Intrusion in Coastal Well Fields", Proceedings of the National Symposium on Ground-Water Hydrology, November, 1967, North Carolina State University, Raleigh, North Carolina, pp. 235-258.

The analogy between the free surface of gravity flow wells and the fresh-salt water interface due to pumping from an artesian aquifer is shown to hold true. Thus, in order to find the combined effects of natural flow and fresh water withdrawal by pumping in coastal fields, the analyses of gravity wells as well as overpumped artesian wells are applied. Salt water intrusion is shown to create a serious problem in coastal well fields which necessitates sound planning and design, especially for the location, depth, and capacity of individual wells. The presented method is an attempt to find the contours of the salt water mounds due to pumping in an artesian aquifer with a certain natural flow. Several assumptions are considered necessary to find a solution for this problem, i.e., evaporation, leakage, etc. are neglected. Although the flow is governed by the Laplace Equation within the fresh water medium, the principle of superposition cannot be valid due to the change of the flow domain. It is suggested, therefore, to superimpose only the individual pressure heads at the upper boundary of the artesian aquifer as well as the individual areas of the pressure head diagrams due to the various effects of the natural flow and the individual wells or their images.

AB-27.   Land, L.F., "Ground-Water Resources of the Riviera Beach Area, Palm Beach County, Florida", USGS/WRD/WRI-77/076, September, 1977, U.S. Geological Survey, Tallahassee, Florida.

The so-called "shallow aquifer" composed chiefly of sand, shells, sandstone, and limestone, is the principal source of freshwater in the Riviera Beach area. The major water-bearing zone consists of cemented layers of sand and shells, about 100 ft thick, in the lower part of the aquifer. The quality of the water in the shallow aquifer is generally suitable for public supply except locally along C-17 Canal where the dissolved solids concentration exceeds 500 mg/l. The configuration of the water table is greatly influenced by Lake Worth, C-17 Canal, West Palm Beach water catchment area, rainfall, and municipal pumpage. The major threat to development of water supplies, and possibly to the continuation of a current withdrawal rate of over 5 mgd, is seawater (Lake Worth), but the combined effects of increased pumpage, reduced recharge resulting from increased land development, and below normal rainfall, have caused seawater to advance inland in the aquifer. Additional supplies could be developed in the west.

AB-28. Martin, P., "Ground-Water Monitoring at Santa Barbara, California, Phase 2--Effects of Pumping on Water Levels and Water Quality in the Santa Barbara Ground-Water Basin", Open File Report 82-366, April, 1982, U.S. Geological Survey, Sacramento, California.

From July, 1978, to January, 1980, water levels declined more than 100 ft in the coastal area of the Santa Barbara ground water basin in southern California. The water level declines are the result of increases in municipal pumping since July, 1978. The pumping, centered in the city less than 1 mile from the coast, has caused water-level declines in the main water-bearing zones to altitudes below sea level. Consequently, the ground water basin is threatened with salt-water intrusion if the present pumpage is maintained or increased. Water quality data suggest that salt-water intrusion has already degraded the water yielded from six coastal wells. Chloride concentrations in the six wells ranged from about 400 to 4,000 mg/l. Municipal supply wells near the coast currently yield water of suitable quality for domestic use. There is, however, no known physical barrier to the continued inland advance of salt water. Management alternatives to control salt-water intrusion in the Santa Barbara area include: (1) decreasing municipal pumping, (2) increasing the quantity of water available for recharge by releasing surplus water to Mission Creek, (3) artificially recharging the basin using injection wells, and (4) locating municipal supply wells farther from the coast and farther apart to minimize drawdown.

AB-29. McCoy, J., "Hydrology of Western Collier County, Florida", Open File Report 72018, 1972, U.S. Geological Survey, Tallahassee, Florida.

Western Collier County, Florida, has a large fresh-water supply potential because of its 54 inches of rainfall annually and its manmade surface flow system, hydraulically connected to a shallow permeable aquifer. However, water problems exist because the rainfall is not evenly distributed throughout the year, causing a threat of salt-water intrusion in the Naples well field during prolonged dry seasons. Contamination of existing and future ground water supplies by man-related activities including urbanization is of concern. The controlled surface-water flow system and the distribution of weirs allow for water control without an excessive lowering of ground water levels. The system also has the potential for recharging the coastal section of the shallow aquifer during the dry seasons. Variable water quality and inadequate flows during the dry season precludes the use of the surface-water flow system as a direct source of municipal water.

AB-30. McCoy, H.J. and Hardee, J., "Groundwater Resources of the Lower Hillsboro Canal Area, Southeastern Florida", Report of Investigation No. 55, 1970, Florida Bureau of Geology, U.S. Geological Survey, Tallahassee, Florida.

The lower Hillsboro Canal area occupies about 60 sq mi of Palm Beach and Broward Counties in southeastern Florida. All potable ground water is obtained from the Biscayne Aquifer. The aquifer extends from the land surface to a depth of about 400 ft and is composed of sand, sandy limestone, shells, and indurated calcareous sand. Municipal well fields of Deerfield Beach and Boca Raton and most of the domestic, irrigation, and industrial wells obtain adequate water supplies from permeable limestone 90 to 130 ft below land surface. Rainfall in the area and induced infiltration from controlled canals provide the recharge to the aquifer. Sea-water intrusion, although a constant threat, has not advanced inland enough to contaminate either municipal well field. Data collection stations are maintained to monitor changes of the salt-water front in the aquifer. Large quantities of water can be withdrawn from the interior part of the area without the threat of salt-water intrusion.

AB-31. Muir, K.S., "Initial Assessment of the Ground-Water Resources in the Monterey Bay Region, California", USGS/WRD/WRI-77/053, August, 1977, U.S. Geological Survey, Menlo Park, California.

Because urban growth has placed an increasing demand on the ground water resources of the Monterey Bay region, an assessment of the ground water conditions was made to aid the development of local and regional plans. Ground water provides 80 percent of the water used in the region, which includes six ground water subbasins. In several of the subbasins, pumpage exceeds safe yield. Existing water quality degradation results from seawater intrusion, septic tank effluent, and irrigation-return water. Potential sources of degradation include municipal sewage disposal, leachates from solid-waste disposal sites, and poor quality connate water. High priority items for future study include location of recharge areas, detection of seawater intrusion, and well monitoring of landfill sites.

AB-32. Muir, K.S., "Seawater Intrusion and Potential Yield of Aquifers in the Soquel-Aptos Area, Santa Cruz County, California", USGS/WRD/WRI-81/020, October, 1980, U.S. Geological Survey, Menlo Park, California.

Seawater has intruded the Purisima Formation in the interval 0-100 ft below sea level in the Soquel-Aptos area. The potential yields of the two principal aquifers in the Soquel-Aptos area are 4,400 ac-ft/yr from the Purisima Formation and 1,500 ac-ft/yr from the Aromas Sand. Pumping from the Purisima Formation, averaging about 5,400 ac-ft/yr since 1970, has caused water levels along the coast to decline below sea level and has allowed seawater to enter the aquifer. Seawater intrusion and ground-water storage could be monitored in all depth zones by expanding the observation-well network.

AB-33.   Park, F.D.R., "Combating Salt-Water Encroachment into the Biscayne Aquifer of Miami, Florida", Bulletin 3, October, 1968, Louisiana Water Resources Research Institute, Louisiana State University, Baton Rouge, Louisiana, pp. 31-56.

Urbanization of Dade County, Florida, and attempts to drain the Everglades have lowered the water table. During periods of drought, the current withdrawal rate of fresh water allows salt-water intrusion. Recharge from rainfall periodically pushes salt water to the sea. It is desirable to artificially hold a high water level to offset intrusion during droughts. Salinity dams, prevention of further primary drainage, and navigation channels are used in salinity control. Little opposition has been encountered in enforcement of the salt barrier line on private lands. The main enforcement difficulties are caused by the public works and mosquito control units. Court decisions subordinate riparian rights to the public need of an uncontaminated fresh water supply. Lesser pollution problems arise from industrial and domestic discharges. An effective answer lies in county-wide sewage collection and disposal.

AB-34.   Rodis, H.G. and Land, L.F., "The Shallow Aquifer--A Prime Freshwater Resource in Eastern Palm Beach County, Florida", USGS/WRD/WRI-76/038, February, 1976, U.S. Geological Survey, Tallahassee, Florida.

A shallow aquifer underlies all of Palm Beach County and is the source of almost all fresh water supplies in the eastern part of the county. It consists of mixtures of sand, shell, sandstone, and limestone. In this area the concentration of dissolved solids in the ground water usually does not exceed 500 mg/l. A section of cavity-riddled limestone and other permeable rocks is located several miles inland and extends north to south almost the entire length of the county. This section yields up to 2,000 gpm (130 liters/sec) of water to large wells and offers an excellent potential for the development of future ground water supplies. Sea-water intrusion into the shallow aquifer is a potential threat to several coastal well fields. The wells nearest the coast are most vulnerable.

AB-35.   Schaefer, F.L. and Walker, R.L., "Saltwater Intrusion into the Old Bridge Aquifer in the Keyport-Union Beach Area of Monmouth County, New Jersey", Water Supply Paper 2184, 1981, U.S. Geological Survey, Washington, D.C.

Lateral saltwater intrusion is occurring in a part of the Old Bridge Aquifer in the vicinity of Keyport and Union Beach Boroughs in Monmouth County, New Jersey. Beginning about 1970, analyses of water samples from five wells tapping the Old Bridge Aquifer show abnormally high chloride concentrations, indicating saltwater intrusion. This saltwater contamination is significant since total annual pumpage from the aquifer by eight major water users in the vicinity of Keyport and Union Beach increased from 3.9 mgd in 1957 to 8.0 mgd in 1976. Water-level measurements in 31 wells tapping the Old Bridge Aquifer were made in 1977 to develop a map of the potentiometric (water-level) surface of the aquifer. This map shows that a cone of depression has developed in the Middlesex-Monmouth County area. The reduction in water levels has caused a reversal in the direction of ground water flow in the Old Bridge Aquifer. Two other potential sources of saltwater contamination are discussed and evaluated: saltwater infiltration through abandoned, unsealed wells, and contamination from excavations for sewer lines.

AB-36. Schiner, G.R., "Ground-Water Conditions in the Cocoa Well Field Area", 1980, U.S. Geological Survey, Orlando Florida.

Increasing chloride content of water pumped by wells of the City of Cocoa has forced the abandonment of some wells and the development of new wells farther to the west. There is increasing concern that salt-water intrusion will eventually contaminate even the newer wells. The City of Cocoa needs information on the relationship between the fresh and salty water in the Floridan Aquifer and ways in which their well field may be protected from future salt water problems. Therefore, a study is being conducted to investigate and monitor the interconnection between the fresh and salty zones of the Floridan Aquifer in the new well field; to evaluate the potential of shallow artesian aquifers; to investigate the feasibility of artificially recharging the Floridan Aquifer in the old well field during the slack pumping season so that good quality water will be available during the season of heavy demand; and to investigate the practicality of various wells to transfer water between the shallow and the Floridan Aquifers. This study will supply the information needed to prevent or control salt-water intrusion in the well field and possibly to reverse it through artificial recharge.

AB-37. Shafer, G.H., "Groundwater Resources of Aransas County, Texas", Report 124, December, 1970, Texas Water Development Board, Austin, Texas.

The aquifers that yield water to wells in Aransas County, Texas, are the Gulf Coast Aquifer, mostly of Pliocene and Pleistocene age, that underlies the entire county; the Pleistocene Barrier Island and beach deposits, which are restricted to the peninsulas; and the Holocene Barrier Island deposits, which are restricted to St. Joseph Island. Almost all ground water needed for public supply and industrial use is obtained from the Gulf Coast Aquifer that underlies Live Oak Peninsula. During 1966, about 1,600 ac-ft of ground water was pumped for all purposes in the county; about 672 ac-ft was for public supply, 342 ac-ft for industrial use, and 580 ac-ft for rural domestic and livestock use. One of the principal factors limiting the development of moderate supplies of ground water is the threat of salt water intrusion from sands underlying the fresh water zone. The most satisfactory method of disposal of salt water is by the use of injection wells, but in 1961 only 500 barrels (0.06 ac-ft) of salt water produced from oil wells in Aransas County was disposed by this method.

AB-38. Smith, C.G., Jr., "Geohydrology of the Shallow Aquifers of Baton Rouge, Louisiana", Bulletin GT-4, October, 1969, Louisiana Water Resources Research Institute, Louisiana State University, Baton Rouge, Louisiana.

The geohydrology of the ten major aquifers in the Baton Rouge, Louisiana, area was studied in order to determine the threat of salt-water intrusion to industrial water supplies. The "400-ft" and "600-ft" sands furnish 22 percent of the annual ground water withdrawal from the aquifers. Salt water has been advancing northward toward the industrial center in the "600-ft" sand as a result of hydraulic gradients created by industrial ground water pumpage. Three miles south of this area, the Baton Rouge Fault obstructs the flow of water from the south side of the fault. Apparently the salt water intruding the industrial wells was trapped in the "600-ft" sand north of the fault and an unknown distance west of the Mississippi River. The "400-600 ft" aquifer complex was not affected to any extent by the Denham Springs fault immediately north of the city. Local pinchouts of the two aquifers obstruct flow north and southwest of the industrial area. The Mississippi River indirectly replenishes the "400-600 ft" complex west of the river through the "University Sand". Removal of the salt water north of the fault will result in natural replenishment with fresh water.

AB-39. Soren, J., "Hydrogeologic Conditions in the Town of Shelter Island, Suffolk County, Long Island, New York", USGS/WRD/WRI-78/024, February, 1978, U.S. Geological Survey, Syosset, New York.

The upper glacial aquifer is the sole source of freshwater for a slowly growing population, currently 2,000 to 8,000 seasonally, on Shelter Island (area 11 square miles), between the north and south forks of eastern Long Island. The aquifer is readily susceptible to lateral infiltration by surrounding saline ground water. Clay beds underlying the aquifer contain saline water, and formations below the clay probably contain saline water. Fresh ground water is mostly soft and low in dissolved solids concentrations; however, several wells near shorelines have yielded salty water from saline-water infiltration. Analyses of well water indicate that contamination of the aquifer by sewage is evident but not severe. The hydrogeologic system is not in equilibrium, and deterioration of water quality is expected to gradually increase as increased pumping of fresh ground water causes further saline-water infiltration and introduces additional effluent to the aquifer from septic tanks and cesspools. Test drilling could help water supply management by determining the extent of the aquifer, and observation wells could provide early detection of saline infiltration.

AB-40.   Stewart, M.T., "Evaluation of Electromagnetic Methods for Rapid Mapping of Salt-Water Interfaces in Coastal Aquifers", Ground Water, Vol. 20, No. 5, September-October, 1982, pp. 538-545.

The study described, conducted in two areas of Florida, demonstrates that the EM conductivity method is useful for rapid, inexpensive ground water surveys where the objective is to locate zones of conductive pore fluids at depths less than 30-40 meters.

AB-41.   Swayze, L.J., "Hydrologic Base for Water Management, Dade County, Florida", 1979, U.S. Geological Survey, Miami, Florida.

Water resources in southeastern Florida, particularly in Dade County, are becoming increasingly influenced by water management practices of the South Florida Water Management District. With the expanding urbanization there is a growing need for fresh water supplies, flood control, and water conservation. Saltwater intrusion has been a major threat to freshwater supplies in coastal Dade County for many years. Numerous municipal, domestic, and industrial supplies have been abandoned because of saltwater intrusion. Agricultural and urban expansion in western Dade County has caused concern of the future water supplies for south Dade County and Everglades National Park. This study is focused on preparing annual reports summarizing the changing hydrologic conditions in Dade County as urbanization continues at a record pace. It will be used to evaluate salt water encroachment in response to changing hydrologic conditions and predict movement of salt water under future conditions. Basic data are collected and analyzed periodically to determine the effects and the needs of water management in the county. Annual reports contain basic data concerning fluctuations of the water table, salt water intrusion and discharges of canals. Water table maps of the county and county well fields are shown for comparison with previous conditions. Graphs and maps showing the distribution of saline water are also included.

AB-42.   Vandenberg, A., "Simultaneous Pumping of Fresh and Salt Water from a Coastal Aquifer", Journal of Hydrology, Vol. 24, No. 1-2, 1975, pp. 37-43.

When fresh water is pumped from an aquifer which is open to a recharging salt water body, care must be taken to avoid the encroachment of salt water into the well. A measure which sometimes may be taken when there is also a demand for filtered salt water is to supplement the fresh-water well by a salt-water well, located between the fresh-water well and the coast, and to pump both wells at comparable rates. This paper analyzes the idealized case where the aquifer is confined, thin, homogeneous and isotropic, semi-infinite in areal extent, and bounded on one side by a straight recharge boundary; it is furthermore assumed that the nonpumping piezometric surface of the aquifer has a steady, uniform slope towards the boundary. As a result of the analysis a relation is established between the distances from each of the wells to the boundary, the ratio between the pumping rates of the two wells, and the maximum pumping rate at which no salt water is induced into the fresh-water well; this relation can be used in the selection of the optimum location of the pair of wells.

AB-43. Walter, G.R., Kidd, R.E. and Lamb, G.M., "Ground-Water Management Techniques for the Control of Salt-Water Encroachment in Gulf-Coast Aquifers", OWRT-B-073-ALA(1), 1979, Office of Water Research and Technology, U.S. Department of the Interior, Washington, D.C.

The fresh ground water resources of the Gulf Shores area of Alabama occur in four principal aquifers. In order of increasing depth, these are designated the Beach Sand Aquifer, the Gulf Shores Aquifer, the 350-Foot Aquifer, and the 500-Foot Aquifer. The 350-Foot Aquifer and the 500-Foot Aquifer contain fresh water at the coastline and may contain fresh water for some distance beneath the Gulf of Mexico. Only in the Beach Sand Aquifer can the cases of salt-water contamination be unequivocally attributed to seawater encroachment. Based on the results of numerical modeling and interpretation of the water chemistry, the cases of contamination in the Gulf Shores Aquifer may be due to leakage of saline water around well bores, leakage from confining beds and movement of pockets of relic salt water within the aquifer rather than lateral seawater encroachment.

AB-44. Walters, K.L., "Reconnaissance of Sea-Water Intrusion Along Coastal Washington, 1966-68", Water Supply Bulletin No. 32, 1971, Washington Department of Ecology, Olympia, Washington.

Rapidly developing areas of the State of Washington are placing large demands on some coastal area ground water supplies that were totally undeveloped a few years ago. The present occurrence and general extent of salt water intrusion into the ground water reservoirs were determined and areas of possible future sea-water intrusion delineated. Collection of data was generally limited to areas within one mile of the shoreline, and the chloride content of ground water was used as the principal criterion for the recognition of sea-water intrusion. Uncontaminated ground water in most coastal areas of Washington contain less than 10 mg/l of chloride. Most wells tapping aquifers intruded by sea water yield water having chloride concentrations of less than about 750 mg/l, but a few have concentrations of more than 5,000 mg/l. Although it has not reached serious proportions, sea-water intrusion has occurred near Olympia, at Tacoma, in northern Kitsap and northeastern Jefferson Counties, and in Island County. Small areas of additional intrusion are expected to develop with increased local pumping, and some intrusion may become more pronounced where it is now only incipient.

AB-45. Wilson, W.E., "Estimated Effects of Projected Ground-Water Withdrawals on Movement of the Saltwater Front in the Floridan

Aquifer, 1976-2000, West-Central Florida", Water Supply Paper 2189, 1982, U.S. Geological Survey, Washington, D.C.

Maps of observed 1976 and simulated year 2000 potentiometric surfaces were used to estimate rates of saltwater encroachment and theoretical predevelopment equilibrium positions of the saltwater-freshwater interface in west-central Florida. The observed saltwater front corresponds closely to a theoretical predevelopment equilibrium position of a saltwater-freshwater interface. The average landward flow rate of the front was computed to be 0.30 foot per day in May, 1976, and 0.36 foot per day in May, 2000. The seaward potentiometric-surface gradient under simulated October, 2000, conditions averaged 8.8×10^{-5} foot per day less than under observed September, 1976, conditions. Net landward movement of the saltwater front in the lower part of the Floridan Aquifer is probably occurring under existing conditions. Pumping during 1976-2000 would probably increase slightly the rate of movement. However, rates are so slow that on a regional basis saltwater encroachment under existing and projected conditions is not presently a threat to the existing freshwater resources.

NONCOASTAL INTRUSION OF SALTWATER

AB-46. Aral, M.M., Sturm, T.W. and Fulford, J.M., "Analysis of the Development of Shallow Groundwater Supplies by Pumping from Ponds", OWRT-A-089-GA(1), February, 1981, Office of Water Research and Technology, U.S. Department of the Interior, Washington, D.C.

Continued pumping from the principal artesian aquifer in coastal Georgia has led to a general decline of the piezometric surface in the Brunswick area. As the ground water level has declined, saline water has intruded into fresh water zones. Since coastal Georgia depends heavily on this ground water resource, its loss would be a severe problem. A shallow, unconfined aquifer overlays the artesian aquifer and receives surface recharge. Two uses for this shallow aquifer have been proposed: (1) recharging the artesian aquifer through connector wells; or (2) using the shallow aquifer as a direct water source through development of large diameter, shallow wells or ponds. This study explores the second alternative through use of a numerical model to study the hydraulics of flow to a pond in an unconfined aquifer. The ground water hydraulics problem is solved by the finite element method for two-dimensional pits and axisymmetric ponds in phreatic aquifers for steady-state and transient cases.

AB-47. Benedict, B.A., Rubin, H. and Means, S.A., "Movement of Salt and Hazardous Contaminants in Groundwater Systems", OWRT-A-047-FLA(1), May, 1983, Office of Water Research and Technology, U.S. Department of the Interior, Washington, D.C.

A method has been developed to significantly reduce

computer resources required to model ground water contaminant movements. The method has been tested by applying it to the stratified flow problem of upward seepage of saltwater into freshwater aquifers overlying semiconfining formations, such as occurs in northeastern Florida. The basic equations of continuity, motion, and solute transport are solved simultaneously after being simplified by the Dupuit assumption and a boundary layer approximation. While the specific example treated salt movement, other contaminants, including hazardous wastes could easily be modeled in the same way in a stratified flow. In addition, if the contaminant did not bring about stratified conditions or occur in a saline-stratified region, the same basic procedure can be utilized by removing contaminant influence on the flow field in the equation of motion.

AB-48. Boggess, D.H., Missimer, T.M. and O'Donnell, T.H., "Saline-Water Intrusion Related to Well Construction in Lee County, Florida", USGS/WRD/WRI-77/061, September, 1977, U.S. Geological Survey, Tallahassee, Florida.

An estimated 30,000 wells have been drilled in Lee County, Florida, since 1900. These wells range in depth from about 10 to 1,240 ft and tap the water table aquifer or one or more of the artesian aquifers underlying the area. Many of these wells are sources of saline-water intrusion into freshwater aquifers. The artesian aquifers in the lower part of the Hawthorn Formation and upper part of the Tampa Limestone and the Suwannee Limestone have high water levels and contain water with dissolved solids concentrations ranging from 1,500 to 2,400 mg/l. Few of the 3,000 wells drilled to these water-bearing units contain sufficient casing to prevent upward leakage into overlying aquifers; intrusion into the upper part of the Hawthorn Formation has resulted in an increase in chloride concentrations from a normal of 80-150 mg/l to 300-2,000 mg/l. Saline-water intrusion into the upper part of the Hawthorn Formation also has occurred as a result of downward leakage of saline water through corroded metal casings of wells drilled in areas adjacent to tidal water bodies. Where this leakage has occurred, the chloride concentration has increased to as much as 9,500 mg/l.

AB-49. Carpenter, A.B. and Darr, J.M., "Relationship Between Groundwater Resources and Energy Production in Southwestern Missouri", OWRT-A-086-MO(1), April, 1978, Office of Water Research and Technology, U.S. Department of the Interior, Washington, D.C.

Seventy farm and municipal water supply wells were sampled in Bates, Barton, and Vernon Counties. The mineralogy of Pennsylvania sandstones and limestones in the area were determined by standard methods. Analysis of the data indicates that the water is in equilibrium with calcite, dolomite, quartz, illite, potash feldspar, and albite. The water is not in equilibrium with kaolinite. The relationships between dissolved sulfur, carbon, iron, pH, and pE indicate that the ground water is not in redox equilibrium. The pronounced fractionation of C12 between methane and associated bicarbonate in the water indicates that the methane is of biogenic origin. The rate and pattern of salt water intrusion in the area have been determined. The movement of selected isochlors between 1966 and 1975 indicates that saline water is migrating in from the northwest and moving in an east-southeastward direction.

AB-50.   Franks, B.J. and Phelps, G.G., "Estimated Drawdowns in the Floridan Aquifer due to Increased Withdrawals, Duval County, Florida", USGS/WRD/WRI-79/069, July, 1979, U.S. Geological Survey, Tallahassee, Florida.

A simplified model of the Floridan aquifer system underlying Duval County consists of two highly permeable water-bearing zones separated by a semiconfining zone of relatively low permeability, all of which are confined above and below by much less permeable strata. At present (1979) only the upper permeable zone is tapped by production wells. Decreases in artesian pressure in the upper permeable zone of the aquifers induce upward leakage of water from the lower permeable zone resulting in reduced pressure in this zone. Thus, saline water from deeper strata can move upward, apparently along fractures and faults into the lower permeable zone.

AB-51.   Rosenau, J.C., et al., "Geology and Ground-Water Resources of Salem County, New Jersey", Special Report No. 33, 1969, New Jersey Department of Conservation and Economic Development, Trenton, New Jersey.

This report discusses the geology pertinent to the development of the water resources of Salem County, New Jersey, the quality of the ground water, and the problem of salt-water intrusion in each aquifer. Use of ground water in Salem County in 1964 averaged 12.28 mgd, or nearly 200 gpd per person. Industrial use was 5.33 mgd and use for public supply was 2.74 mgd. Irrigation, which has become increasingly important in recent years, accounted for 2.14 mgd. Average total summer use of ground water--approximately 17.7 mgd--was considerably higher than the yearly average. The Potomac Group and Raritan and Magothy Formations contain the most productive aquifers in Salem County. These aquifers are hydraulically connected in places and may be considered to form a single hydrologic unit or aquifer system. Reported yields of wells tapping this system range up to 860 gpm. A decline in water levels in this aquifer system has resulted from heavy pumpage. The aquifer in the Wenonah Formation and Mount Laurel Sand is the second most highly used aquifer in Salem County and is an important source of water for future development. Reported yields of wells tapping this aquifer range up to 507 gpm. Salt-water intrusion into the aquifer in the vicinity of the City of Salem is indicated by chloride concentrations of up to 396 gpm. The aquifer in the Vincentown Formation is an important aquifer in part of Salem County. It is capable of supplying considerably more water than is now being pumped and, hence, is important for future ground water development. Reported yields of wells tapping this aquifer range up to 270 gpm.

AB-52. Shedlock, R.J., "Saline Water at the Base of the Glacial-Outwash Aquifer Near Vincennes, Knox County, Indiana", USGS/WRD/WRI-80/067, August, 1980, U.S. Geological Survey, Indianapolis, Indiana.

A plume of saline water at the base of the glacial outwash aquifer near Vincennes, Indiana, has been drawn into the municipal well field. However, the average chloride concentration of the municipal water, 30 $\pm$ 15 mg/l, did not change significantly from 1976 to 1979. The plume, an elongated lens approximately 6,500 ft long by 1,500 ft wide, is 4 ft thick near the well field, and the chloride concentration of the water is 3,500 mg/l. The saline water seems to be entering the outwash aquifer through bedrock fractures. The fractures probably intersect unplugged intervals of the abandoned oil wells that convey saline water from bedrock aquifers at unknown depths. Digital model analysis indicates that doubling the 1979 pumping rate in the well field would cause water level declines of generally less than 8 ft. Solute-transport model analyses indicate that the chloride concentration of the well field water would be less than 250 mg/l for a saline-water intrusion rate ten times the model-calibrated rate.

AB-53. Wilmoth, B.M., "Development of Fresh Ground Water Near Salt Water in West Virginia", Ground Water, Vol. 13, No. 1, January-February, 1975, pp. 25-32.

Adequate well fields have recently been successfully constructed just above shallow salt water in bedrock aquifers at Hattie in Calhoun County, near Madison in Boone County, near Southside in Mason County, and at Prichard in Wayne County. In all of these areas of successful construction, the essential information for initial test drilling was obtained by detailed hydrogeologic work at the prospective sites. Most important was the determination of the maximum depth of fresh water, well spacing and pumping rates.

OIL-GAS FIELD ACTIVITIES

AB-54. Cities Service Oil Co. vs. Merritt (Damages for the Pollution of Subterranean Waters), 332 P2D 677-688 (OKLA. 1958).

Plaintiff landowner sued defendant petroleum companies for actual and punitive damages resulting from the salt water pollution of subterranean waters underlying plaintiff's water wells. The wells became permanently unpotable. Defendants alleged error by the trial court in: (1) not requiring plaintiff to prove negligence, (2) awarding excessive actual damages, and (3) awarding punitive damages improperly. The Supreme Court of Oklahoma held that: (1) the basis of liability for injury to property by oil pollution and salt water intrusion is either negligence or nuisance, (2) liability for a nuisance does not require proof of negligence since negligence is not an essential element of a cause of action for nuisance, (3) oil companies which are joint tortfeasors are each accountable for the entire amount of damages. The court ruled that the plaintiff was entitled to damages measured by the difference in land value prior to and just after the injury. Plaintiff was held entitled to reimbursement for emergency expenditures made to minimize damages. Punitive damages were held allowable because defendants had intentionally permitted salt water to escape into nearby creeks. The punitive damages awarded, however, were held to be excessive.

AB-55. Donaldson, E.C., "Environmental Aspects of Enhanced Oil Recovery", Energy Sources, Vol. 4, No. 3, 1979, pp. 213-229.

Oil displacement by injection of brines (the waterflood process) entails the surface processing of large quantities of saline waters, creating the potential for pollution of surface waters and shallow freshwater aquifers. Improved waterflooding techniques involve the introduction of surface active agents (detergents), caustics, and organic polymer compounds (particularly polyacrylamide), thus increasing the potential for pollution. It is increased further by the micellar-polymer process, since the micellar solution is a mixture of surfactant, oil, water, and a cosurfactant (usually an alcohol). Enhancement of oil production by thermal methods adds another dimension to the environmental hazards through air pollution from steam generators and the produced fluids. Other field operations, such as drilling and renovation of wells, create local environmental hazards, but these are becoming regional concerns as the search for new oil and the reactivation of old oil fields accelerates. This paper presents the potential pollution hazards of the enhanced oil-recovery processes and the environmental surveillance program being instituted by the Department of Energy (DOE).

AB-56. General Crude Oil Co. vs. Aiken (Groundwater Pollution by Saline Water from Oil Drilling Operations as Creating Permanent Damages), 335 S.W. 2D 229-235 (Civ. App. Tex 1960).

Plaintiff ranchowner sought to recover damages for the permanent decrease in land value from operations of defendant oil company. Defendant was the owner of an oil and gas lease from plaintiff and operated three oil wells on plaintiff's land. Plaintiff contended that salt water was produced from the oil wells and was placed into evaporating tanks which were inadequate to handle the large volume of water. Plaintiff contended that as a result of defendant's negligent operation, plaintiff's fresh water spring and entire ground water supply was polluted by salt water. The trial court awarded damages to plaintiff for the permanent injury to his land. Defendant appealed claiming that the damages were not permanent in nature and the damages incorrectly included a reduction in the value of the mineral lease. The Court of Civil Appeals of Texas reversed and remanded the case, holding that the instructions

concerning the amount of damages incorrectly included the
decline in oil production. The court, however, did find ample
evidence of permanent damages from salt water intrusion.
Injury to land may be permanent in law without being perpetual.

AB-57. Meier, V., "Oil Well Operation and Salt-Water Intrusion: Policy
Implications", Industrial Wastes, Vol. 24, No. 2, March-April, 1978, pp.
42-45.

Regulations and methods to prevent oil, salt water, and
waste water intrusion into ground water during the drilling of oil
wells are reviewed. Waste water from oil drilling is generally
contained in earthen pits which may be lined with a gel and
polymer, plastic, or steel. The volume of waste water is
evaporated, pumped into a well annulus, injected into salt water
disposal wells, buried in trenches, or applied to land. Defoamed
water is settled in a series of pits before it is discharged into
streams. Casings for wells are drilled 30-100 ft below the fresh
water horizon and lined with cement that can be returned to the
surface. Intermixing of oil, gas, fresh, or salt water strata above
the source of oil is prevented by cementing and casing. Fuel
tanks are contained by dikes; waste oil is collected in ditches and
dikes. Tank overflow and backwash are channeled to plastic-
lined pits for evaporation or injection well disposal. Compressor
stations or cooling towers treat the gas produced during drilling
before it is pumped into pipelines. Oil and grease in cooling
streams are skimmed to 10 mg/l before discharge; water
discharged from drilling operations has an oil content not greater
than 50 mg/l.

AB-58. Pickens vs. Harrison (Damages to Irrigation Well Caused by Salt Water
Pollution from Oil Well), 246 S.W. 2D 316-319 (Civ. App. Tex. 1952).

Plaintiff landowners sued defendant oil well owner to
recover damages to plaintiff's lands caused by salt water
intrusion into irrigation water. Plaintiffs contended that
defendant had negligently allowed salt water from his oil well
operations to run onto their lands, thereby polluting their
irrigation well and ruining their rice crops. Defendant denied
liability and contended that any pollution was partially caused by
other neighboring oil wells. Defendant also sought to introduce a
letter from a chemist to the plaintiff's allegedly stating that
analysis of the salt content of the irrigation wells showed that
the water was not polluted. In affirming a jury verdict and
judgment for plaintiffs, the Court of Civil Appeals of Galveston,
Texas, held that the evidence produced by plaintiffs was
sufficient to support a verdict for them. In a suit for damages
caused by water pollution, a letter from a non-party third person
relating to chemical analysis of the water is hearsay evidence
and is inadmissible. According to evidence related to the
geography and soil condition of the property, there was ample
support for a finding that the water pollution was caused solely
by defendant's well.

AB-59. Schlichtrull vs. Mellon-Pollock Oil Co. (Negligent Drilling of Oil Well

Alleged to Have Caused Saline Water Intrusion), 301 PA 553, 152 A 829-931 (1930).

Plaintiff farmowner sued defendant oil well lessee for damages caused by pollution of plaintiff's well. Defendant had encountered salt water in drilling; and because defendant had failed to drive casings to prevent intrusion of the salt water into the fresh water strata lying above, the salt water seeped into plaintiff's well. The court noted that without negligence no liability inures for interference with subterranean waters. However, defendant was liable because it had failed to exercise due care. Defendant asserted that the oil well was to be abandoned in six months and that, therefore, the injury was temporary. Plaintiff asserted that the injury was permanent, since defendant had not proven that plugging the well would prevent salt water intrusion or restore the property to its original condition. Holding that the injury was permanent, the court affirmed the trial court's measure of damages based upon the diminution in market value of plaintiff's tract.

AB-60. Shafer, G.H., "Ground-Water Resources of Nueces and San Patricio Counties, Texas", Report 73, May, 1968, Texas Water Development Board, Austin, Texas.

In 1964, 15.6 mgd was pumped, of which public supply used 2.3; irrigation, 8.2; industrial, 2.1; and domestic and stock, 3.0 mgd. The coefficient of transmissibility ranges from 1500 to 24,000 gpd/ft. Further development of a few mgd may be possible without depletion by salt-water intrusion in northwest San Patricio County, where yields could be 1700 gpm. A few million acre-ft might be withdrawn economically from storage. Large quantities of moderately saline water are available. Injection wells are considered the best way to dispose of oil-well salt water, but only 23.9 percent of the total quantity in Nueces County and nine percent in San Patricio County is injected.

AB-61. Wendtlandt vs. National Co-Operative Refinery Association (Salt Water Intrusion Resulting from Oil Drilling Operations), 215 P. 2D 209-215 (Kan. 1950).

Plaintiff landowner sought to recover actual and punitive damages for permanent injury to her land caused by salt water from defendant's oil drilling operations. Defendant operated several oil wells on plaintiff's land pursuant to a lease arrangement. The oil wells produced over 500 barrels of salt water per month. Defendant disposed of this by pumping it into a large pit with a capacity of 3,500 barrels. Defendant knew the salt water was seeping into the soil, yet took no corrective action. Portions of plaintiff's land became polluted by salt water and useless for agricultural purposes. Plaintiff alleged that the value of her land had been permanently decreased by $6000. Plaintiff was awarded $1092 in permanent damages and $5000 in punitive damages. Defendant appealed, contending that the evidence was insufficient to warrant punitive damages. The Supreme Court to Kansas affirmed the judgment on condition that the punitive damages be reduced to $2500. The court held that punitive damages are permissible where a defendant intentionally empties salt water on a plaintiff's land, knowing it would cause serious damage.

IRRIGATION EFFECTS

AB-62.   Balsters, R.G. and Anderson, C., "Water Quality Effects Associated with Irrigation", Kansas Water News, Vol. 22, No. 1 and 2, Winter, 1979, pp. 14-22.

Ground water is used to irrigate 95 percent of 3 million acres of Kansas cropland. Irrigation may increase salt concentration and add sediments, fertilizers, and pesticides to runoff and return water making the water unsuitable for agricultural, industrial, and domestic uses. In addition, overpumping an aquifer may cause salt water intrusion from an adjacent brine aquifer. The geochemistry of the various water quality constituents in irrigation water is discussed. Factors that alter irrigation water quality are evapotranspiration, erosion, ion exchange, leaching, filtration, consumptive use by crops, chemical composition of the soil, irrigation methods, crop variety, and the initial quality and quantity of the irrigation water. Thus, degradation of irrigation water is highly variable and site specific.

AB-63.   Helweg, O.J., "Nonstructural Approach to Control Salt Accumulation in Ground Water", Proceedings of the Third National Water Well Association-EPA National Ground Water Quality Symposium, Las Vegas, Nevada, September 15-17, 1976, pp. 51-57.

This paper presents several strategies that offer a possibility of controlling the salt buildup in areas of pollution occurring in ground water from normal irrigation practices. Instead of applying irradiation ground water near the site of the well, the ASTRAN Method transfers it downstream to be applied on land where the ground water is a lower quality thereby controlling the increase in salt concentration. Instead of preventing seepage loss in delivery canals, the percolating water is used to maintain ground water quality.

AB-64.   Pincock, M.G., "Assessing Impacts of Declining Water Quality on Gross Value Output of Agriculture, Case Study", Water Resources Research, Vol. 5, No. 1, February, 1969, pp. 1-12.

A methodology for assessing salinity damages is still in the development stage; this study develops economic analyses of the effects of different levels of water quality on output and income for irrigated agriculture; first, the procedure for estimating physical crop losses due to salinity was summarized from literature; second, published enterprise and farm budgets were adjusted to show damages that might be expected for a specific irrigation project.

FRESHWATER (INJECTION AND SURFACE WATER PRESSURE-HEAD) AS A BARRIER TO SEA-WATER

AB-65. Bruington, A.E., "The Amelioration or Prevention of Salt-Water Intrusion in Aquifers--Experience in Los Angeles County, California", Louisiana Water Resources Research Institute Bulletin 3, October, 1968, Louisiana State University, Baton Rouge, Louisiana, pp. 153-168.

Underground storage from winter storms is used in Los Angeles County, California, with the pressure-ridge method for recharge of the aquifers for deterring saline intrusion. A line of injection wells creates a pressure field much like a continuous curtain wall. Waterlogging has been controlled in low areas by seaward extraction along with pressure-ridge injection. Injected water must be of high quality to prevent clogging of wells, increase well life, and reduce well-cleaning costs. Observation wells are needed at critical points along the barrier line to monitor the ground water elevation. The location of the barrier line has allowed a certain amount of salt water to be cut off, or trapped behind the barrier. The cost of injected water for 1966, $13.00 per acre-foot, did not include the cost of supply water, amortized capital outlay per well, pipelines, and other appurtenances. The uncompleted pressure-ridge system is working as planned.

AB-66. Bruington, A.E., "Control of Sea-Water Intrusion in Ground-Water Aquifer", Ground Water, Vol. 7, No. 3, May-June, 1969, pp. 9-14.

Seawater intrusion into ground water reservoirs occurs when permeable formations outcrop into a body of seawater and when there is a landward gradient; intrusion can be controlled by reducing pumping, by increasing supply, or by forming some type of barrier. A pressure barrier has been operated in Los Angeles County, California; special facilities are needed; and the costs of operations are high.

AB-67. Cherry, R.N., "Hydrology of Clearwater-Dunedin, Florida", 1980, U.S. Geological Survey, Tampa, Florida.

The Cities of Clearwater and Dunedin, Florida, are faced with the possibility of having to abandon the local aquifer as a source of fifty percent of their supply because of increases in salinity in their production wells. To determine the most efficient arrangement of wells and rates of pumping to obtain the greatest amount of water without causing further salt water intrusion a project was conducted to: (1) perform pumping tests to determine the aquifer characteristics; (2) test drill in the area to determine the location of the salt-fresh water interface; and (3) sample existing wells to determine the extent of salt water encroachment.

AB-68. Jeffers, E.L. et al., "Water Reclamation and Automated Water Quality Monitoring", EPA-600/2-82-043, March, 1982, U.S. Environmental

Protection Agency, Municipal Environmental Research Laboratory, Cincinnati, Ohio.

The Santa Clara Valley Water District owns and operates a water reclamation facility located in the Palo Alto Baylands area in northern California. The purpose of the facility is to provide reclaimed water suitable for injection into the ground water, thereby providing a salt water intrusion barrier and, secondarily, to provide a research facility for various ongoing projects. The project results reported here involved using the NASA Ames Research Center's Water Monitor System to collect data on the water reclamation process train. The Water Monitor System is a continuous on-line water quality monitoring system which automatically measures fourteen water quality parameters in addition to nine trace halocarbons.

AB-69.   Jenks, J.H. and Harrison, B.L., "Multi-feature Reclamation Project Accomplishes Multi-Objectives", Water and Wastes Engineering, Vol. 14, No. 11, November, 1977, pp. 19-24, 30.

The Santa Clara Valley Water District's Palo Alto Reclamation Facility uses reclaimed wastewater treatment facilities to preserve ground water from salt water intrusion and provides an alternate water source for a number of direct uses. It accomplishes a reduction in discharge of wastewater to the lower San Francisco Bay and demonstrates the efficacy of reclaimed wastewater utilization for supplementing the usable ground water supply. Distinctive features include the use of cooling aerators to strip ammonia for N reduction; the use of ozonation before carbon adsorption for disinfection and for preconditioning for organics removal; provision for filtration before and after adsorption to insure complete removal of particulates; computer control and data logging for both the reclamation plant and the injection-extraction well field; flexibility of operation for various treatments; a doublet injection-extraction well system to form a salt water intrusion barrier and control against inland migration of reclaimed water to usable wells; the institution of a surveillance, monitoring, and testing program; and an outstanding cooperative effort between sewering agencies and the water district.

AB-70.   Kashef, A.I., "Control of Salt-Water Intrusion by Recharge Wells", Journal of the Irrigation and Drainage Division, American Society of Civil Engineers, Vol. 102, No. IR4, December, 1976, pp. 445-457.

Various effects can be obtained by the arrangement of recharge well batteries in different patterns on the degree of salt-water retardation in coastal confined aquifers. The batteries are located at various distances parallel to the shoreline; each battery consists of equally spaced recharge wells of a certain number and all wells start operation at the same time. Theoretical formulas have been developed; available well formulas of water flow from recharge wells are considered without modifications. These formulas assume complete penetration of isotropic aquifers. A line source rather than actual well diameter is also assumed. In all investigated cases, a special datum is selected in such a way that a dimensionless total head would be equal to unity at the boundary between entirely freshwater and salt-water intruded zones. Equipotential intensities are determined for all cases at the nodes of a selected grid system. The resulting set of curves summarizes all cases and may be useful in management problems. An example is solved which illustrates the use of the curves.

AB-71. Lindenbergh, P.C., "Effective Use and Conservation of Groundwater", Freshwater, Council of Europe, Nature and Environment Series, No. 2, 1968, pp. 71-106.

Ground water is one portion of the earth's hydrologic cycle. It is important to investigate development of ground water for water supply in the most economical way. It is to be expected that the demand for ground water will continue to increase. At present, overdraft already occurs in many ground water basins. Until overdrafts are reduced to safe yields in ground water basins, permanent damage or depletion of ground water supplies must be anticipated (restrict demand). Artificial recharge is useful for reducing or terminating overdraft and may be defined as augmenting the natural replenishment of ground water storage and, where necessary, preventing salt-water intrusion. National and international ground water research programs are necessary to anticipate future problems and propose technical and economical solutions before ground water projects are designed and planned.

AB-72. Montanari, F.W. and Waterman, W.G., "Protecting Long Island Aquifers Against Salt-Water Intrusion", Bulletin 3, October, 1968, Louisiana Water Resources Research Institute, Louisiana State University, Baton Rouge, Louisiana, pp. 59-86.

Salt water intrusion in New York involves only the perimeter of Long Island. The problem is compounded by direct contact between the aquifers and the ocean. All Long Island ground water comes from infiltration with approximately fifty percent absorption into the aquifers. Since 1933, legislation has been enacted to regulate the ground water extraction and to license well drillers. The main causes of salt-water intrusion are overpumpage and a reduction in availability of recharge water as a result of urbanization. Projects undertaken to relieve overpumpage are: (1) reservoirs to collect surface water, (2) acquiring pumping rights of sparsely populated areas, and (3) construction of a desalination plant. Plans are under way to salvage 97 mgd of presently wasted sewage water for injection. Presently, Long Island is using recharge basins with a capacity of 57 million gallons. Present efforts, such as recharging, are being offset by wasteful sewage discharge.

AB-73. Muir, K.S., "Seawater Intrusion, Ground-Water Pumpage, Ground-Water Yield and Artificial Recharge of the Pajaro Valley Area, Santa Cruz and Monterey Counties, California", USGS/WRD-75-009, October, 1974, U.S. Geological Survey, Menlo Park, California.

The Pajaro Valley area, California, covering about 120 sq mi (310 sq km), extends from the southern part of Santa Cruz County to several miles south of the county line into Monterey County. Seawater intrusion is occurring in the Pajaro Valley area from several miles north to several miles south of the mouth of the Pajaro River and in a small area about four mi (6.5 km) north of the river. The intrusion extends inland about one mi (1.5 km). Two water-bearing zones are being intruded--the depth intervals 100-200 ft (30-60 m) and 300-600 ft (90-180 m). Ground water pumpage averaged 49,100 acre-ft/yr for the nine-year period 1963-71. The long term ground water yield of the Pajaro Valley is about 44,000 acre-ft/yr. Artificial recharge can be effected through modified streambeds at infiltration rates as high as 3 ft/d (0.9 m/d). Injection wells may have recharge capabilities of as much as 500 gal/min.

AB-74. Owen, L.W., "The Challenge of Water Management: Orange County Water District, California", Bulletin 3, October, 1968, Louisiana Water Resources Research Institute, Louisiana State University, Baton Rouge, Louisiana, pp. 105-125.

The technical aspects of effective basin management of Orange County, California, water supplies are described. Reliance has been placed on both surface water and ground water to meet county needs. Surface water is obtained from the Colorado and Santa Ana Rivers and direct precipitation. Ground water comes from the south coastal basin. Feasible use of these sources is dependent on safe withdrawal of large volumes of ground water during periods of surface water shortage and the ability to sink large quantities of water into the basin when surplus surface water is available. Salt-water intrusion will be controlled by a hydraulic barrier of injection wells. The major problem encountered in injecting reclaimed and surface water is the quality of the available water. Blending of surface water and ground water provides a desirable composite water source. Computer models have greatly facilitated working knowledge of the ground water basin operation.

AB-75. Peters, J.A. and Rose, J.L., "Water Conservation by Reclamation and Recharge", Journal of the Sanitary Engineering Division, American Society of Civil Engineers, Vol. 94, No. SA4, August, 1968, pp. 625-638.

Nassau County, New York, is planning to reclaim water from the effluent of its water-pollution-control plants, and inject this water into aquifers furnishing most of the public water supplies in the county. The injection will create a hydraulic barrier, preventing salt-water intrusion, and will allow increased withdrawals from existing water supply wells. This paper describes the relationship of the Nassau County program to other salt-water barrier and advanced waste treatment projects, and examines water-quality standards for recharging and process selection to meet these standards: demonstration facilities consisting of a 400-gpm renovation plant providing coagulation, filtration, and carbon adsorption, and a water injection plant with redevelopment facilities and observation wells are also described.

AB-76. Wedding, J.J., Fong, M.A. and Crafts, C.M., "Use of Reclaimed Wastewater to Operate a Seawater Intrusion Control Barrier", Proceedings of Water Reuse Symposium, Mar. 25-30, 1979, Washington, D.C., Vol. 1, 1979, American Water Works Association, Research Foundation, Denver, Colorado, pp. 639-662.

Reclamation of wastewater from the Oxnard Wastewater Treatment Plant to create a seawater intrusion control barrier and to irrigate agricultural sites has been evaluated as a viable reuse alternative. A process train which includes multi-media filtration, reverse osmosis (or reverse osmosis in conjunction with a mineral source control program) and chlorination will be utilized to produce reclaimed water of suitable quality. Agricultural irrigation with reclaimed water provides for a reliable and abundant alternate water supply for the Oxnard Plain. This alternative is technically feasible and would have the beneficial effect of reducing the overdraft situation. The paper discusses the background of the problem treatment processes, reclaimed water characteristics, injection/extraction well system, and other aspects of the wastewater reclamation and reuse plan.

INSTITUTIONAL MEASURES

AB-77. Bookman, M., "Legal and Economic Aspects of Salt-Water Encroachment into Coastal Aquifers", Bulletin 3, October, 1968, Louisiana Water Resources Research Institute, Louisiana State University, Baton Rouge, Louisiana, pp. 159-192.

An integral part of salt-water intrusion prevention is control and reduction of ground water production, which, in Los Angeles County, California, required court action. Legislative action was also necessary to empower local water districts to construct, operate, and finance salt-water prevention facilities. At present, seventy percent of the water is imported. Implementation of controls was a three-step process: (1) sharing of imported river water with upstream owners; (2) injunctions against coastal and central basin pumpers; and (3) establishment of a salt-water barrier. Barrier financing came through advalorem taxes and district general funds. Other legislation provided for financing enforcement. Salt-water prevention in the west coast basin is administered by six state and local agencies and one citizen's association. This arrangement has proved successful and economical. The control of pumping and reduction of ground water use by pricing is not sufficient to halt salt-water intrusion. Court actions and their consequences are an important economic factor.

AB-78. Fractor, D.T., "Property Rights Approach to Groundwater Management", Dissertation Abstracts International, Vol. 43, No. 9, 1983, p. 3064.

This study presents an institutional, theoretical, and applied analysis of ground water allocation in the arid West, with special reference to California. Ground water is by nature a common property resource. In the absence of formal institutions regulating its use, ground water is available according to a "rule of capture", since there are no exclusionary private property rights to the resource. Stated simply, each ground water user within a basin inflicts costs on other pumpers by his own pumping. Ignoring these costs results in pumping at a higher rate than that which maximizes aggregate discounted returns from ground water. Given an abundance of ground water, these external costs may be modest. Today, however, problems with drawdown, land subsidence, salt water intrusion, and increasing pumping costs suggest pressure for institutional reform.

AB-79. Helwig, J.G., "Salt Water Intrusion in Florida (Legal Approaches to Prevention and Control)", 1969, Eastern Water Law Center, University of Florida, Gainesville, Florida.

Sea water, which is heavier than fresh water because of dissolved salts, moves into fresh water supplies in coastal areas if the fresh water level becomes too low. Invading salt water may extend upstream where tidal waters have access to the stream and the stream flow is not great enough to sweep the salt water out. If the fresh water aquifer is open to the sea, lowering the freshwater level permits direct inflow of salt water. Also, residual salt water trapped beneath impervious rock layers may contaminate higher fresh water aquifers when improper well drilling techniques or equipment allow vertical mixing of the aquifers. Salt water intrusion in Florida results from increased water use, excessive drainage, lack of protection against tidal waters in canals, improper drilling techniques, and waste. Corrective legal measures include artesian well capping statutes to prevent waste, implementation of salt water barrier lines in tidal channels, and licensing and regulation of well drilling. Florida has either enforcement or statutory problems in each of these areas.

AB-80. Hodges, F.D.C., "Water Legislation and Institutions--Their Role in Water Resources Development and Management", Water International, Vol. 2, No. 3, September, 1977, pp. 3-9.

As countries develop, using their available water to an increasing extent, it is a sad fact that water resources legislation tends to lag behind the need for it. Though conditions differ greatly from country to country, thus necessitating respective differences in water laws and institutions, one common factor--the best interests of the inhabitants--should be the basis for any nation's legislation. This paper provides an overview of existing systems of laws relating to water use from both surface and underground sources. Three broad categories of existing legal systems are discussed: riparian rights, prior appropriation, and administrative disposition of water-use rights. Further comments are made on the subjects of priorities for use, surface-ground water relationships, abatement of pollution, land reform, overpumping and salt water intrusion, and governmental powers in water emergencies. The paper advocates a democratic and inter-disciplinary approach to worldwide water resources planning in the future.

AB-81.   Jackson, W.T. and Paterson, A.M., "The Sacramento-San Joaquin Delta and the Evolution and Implementation of Water Policy: An Historical Perspective", W77-10101, June, 1977, Office of Water Research and Technology, U.S. Department of the Interior, Washington, D.C.

An historical review is presented of the attempts to control the encroachment of salinity from San Francisco Bay into the Delta of the Sacramento and San Joaquin Rivers from 1920 to the present. Two basic solutions to the problem were: the physical separation of salt water from fresh water by means of a barrier at some point downstream from the Delta, or the release of sufficient water from upstream storage reservoirs to drive back tidal salinity. Water policy in the Delta has always been intimately linked with plans for the control and redistribution of northern California's water supply, particularly the Central Valley Project and State Water Project. In the late 1960's, environmental considerations involving the use of water for fish and wildlife protection became increasingly significant determinants of water policy in the Delta.

AB-82.   Maloney, F.E., "Laws of Florida Governing Water Use", Journal of American Water Works Association, Vol. 47, No. 5, May, 1955, pp. 440-446.

Past and potential increases in irrigational and industrial demands for water in Florida, combined with increasing problems of salt-water intrusion and domestic uses, evidence a need for a reevaluation of Florida's water laws. Most of Florida's water law is based upon case rulings. The water use rule as to watercourses is the reasonable use rule subject to a similar right of use by all riparian owners. Thus the question is whether diversion for irrigation or industrial use would be reasonable. The two faults of this rule are its indefiniteness and its failure to provide for diversion for use on non-riparian land. The law of reasonable use has been applied to ground water following a definite channel. Percolating waters are governed by the doctrine of correlative use or the doctrine of reasonable use. None of the above standards are clearly established by case law or sufficient to give a satisfactory criteria to control water use. A possible answer is to replace the case law with a comprehensive water code. The need for water planning becomes more apparent every day and the Governor's Citizen Committee on Water Resources is a constructive first step.

AB-83.   McDonald and Grefe, Inc., "Institutional and Financial Alternatives and Recommendations: AMBAG Section 208 Water Quality Management Plan", May, 1978, San Francisco, California.

This report describes the institutional mechanisms, intergovernmental arrangements and sources of financing that are recommended to implement the Association of Monterey Bay Area Governments Water Quality Management Plan for Monterey and Santa Cruz Counties. Recommendations are presented for institutional responses to a total of eight case studies (lake maintenance, septic system maintenance, control of erosion and sedimentation, control of nitrate pollution of ground water, preventing salt water intrusion, preserving ground water recharge areas, and prevention of pollution from sanitary land-filled leachate). Recommendations are also made for a continuation of the Section 208 Water Quality Management Planning process. The institutional arrangements feature maximum reliance on existing powers of existing local governments. A newly authorized governmental mechanism in California--the On-site Wastewater Disposal Zone (OWDZ)--is recommended for management of septic systems.

AB-84. Smith, S.C., "The Role of the Public District in the Integrated Management of Ground and Surface Water", Economics and Public Policy in Water Resource Development, edited by S.C. Smith and E.N. Castle, Iowa State University Press, Ames, Iowa, 1964, pp. 341-352.

In this article, the Public District is discussed as a public agency which is used in organizing integrated management of ground and surface water. Through the organizational process, integration can be achieved among both internal and external interests if the District has the legal, political and economic responsibility to perform this function. The Public District is therefore an agency for collective action. This agency can function both where there is no state statutory administration of water rights and in appropriative states, especially in enacting programs such as combating physical uncertainties due to cyclic fluctuations, salt-water intrusion or compaction. Some management practices of the District may affect the legal status of water rights. The District can act to settle conflicts of interests internally by integrating the management of ground and surface water. The District can serve to relate repayment of a current investment in water utilization facilities to economic benefits accruing to the area over time. Finally, the District can represent internal interests within the District on a common basis to external interests in water management.

MODELING: COASTAL SEAWATER INTRUSION--PHYSICAL

AB-85. Cahill, J.M., "Hydraulic Sand-Model Study of the Cyclic Flow of Salt Water in a Coastal Aquifer", Professional Paper 575-B, 1967, U.S. Geological Survey, Phoenix, Arizona, pp. B240-244.

Model studies demonstrate visually that "cyclic flow" takes place in the denser fluid when fresh water moves seaward over intruding ocean water. When tidal fluctuations are simulated in the model by alternately raising and lowering the level of the free body of salt water, additional mixing of the two miscible fluids occurs and additional salt water is transported into the fresh-water region. The fluid movement was observed by following colored tracers that were injected into both fluids. These studies indicate that the degree of salt-water intrusion is controlled primarily by the flow of the fresh water rather than by tidal effects.

AB-86. Christensen, B.A. and Evans, Jr., A.J., "A Physical Model for Prediction and Control of Saltwater Intrusion in the Floridan Aquifer", WRRC-PUB-27, January 29, 1974, Water Resources Research Center, University of Florida, Gainesville, Florida.

Continuing development of the coastline zone in the middle Gulf area of Florida is increasing the demand for ground water supplies, and in turn increasing the probability of salt-water intrusion. Methods must be developed to make long range predictions on the effects of increased demands on the Floridan aquifer. The section selected for study lies in a line from the Gulf coast near Tarpon Springs to a point near Dade City and passes through the Eldridge-Wilde Well Field. A Hele-Shaw Model has been built for this section, and all pertinent geological and hydrological features of the area are included. Steady-state characteristics of the aquifer system have been considered. In particular, the long-term effects due to pumping and artificial recharge were examined.

AB-87. Collins, M.A., "Ground-Surface Water Interaction in the Long Island Aquifer System", Water Resources Bulletin, Vol. 8, No. 6, December, 1972, pp. 1253-1258.

Steady state experimental studies with a viscous analog of the aquifer system in central Long Island, New York, have shown there to be significant interaction between surface accretion, stream base flow, well recharge, and the degree of salt water intrusion. Reductions in accretion are found to cause a proportionately larger decrease in stream base flow. The degree of intrusion is found to be related to the distribution of accretion and well recharge between stream base flow and submarine flow to the sea.

AB-88. Kashef, A.I., "Model Studies of Salt Water Intrusion", Water Resources Bulletin, Vol. 6, No. 6, December, 1970, pp. 944-967.

A summary of the history of viscous flow model studies of salt water intrusion is given with emphasis on Hele-Shaw Models. The theoretical approach proposed by the writer was checked by using a Hele-Shaw Model. In addition to the study of the steady state cases of fresh water flow, three transient conditions have been experimentally studied.

AB-89. Kritz, G.J., "Analog Modeling to Determine the Fresh Water Availability on the Outer Banks of North Carolina", Report No. 64, April, 1972, North Carolina Water Resources Research Institute, North Carolina State University, Raleigh, North Carolina.

The major purposes of the investigation were to determine the suitability of a Hele-Shaw Model for studying potable water availability on long oceanic islands, to predict maximum safe pumping rates that avoid dangerous salt water pollution of the fresh water lens, and to predict the approximate location of the fresh water-salt water interface on the Outer Banks of North Carolina under various conditions of infiltration, island characteristics, and pumping.

AB-90. Mualem, Y. and Bear, J., "Shape of the Interface in Steady Flow in a Stratified Aquifer", Water Resources Research, Vol. 10, No. 6, December, 1974, pp. 1207-1215.

This work deals with the shape of the interface in a coastal aquifer in which a thin horizontal semipervious layer is present. Field observations and laboratory experiments have shown that under these conditions a discontinuity in the shape of the interface occurs such that a freshwater region exists under the semipervious layer while immediately above it, saline (or mixed) water is present in the aquifer. The paper presents an approximate solution for the shape of the interface below the semipervious layer and for the extent of the freshwater region above it under steady state conditions. The solution is based on the Dupuit Assumption and on a linearization of part of the flow equations. Laboratory experiments on a Hele-Shaw Model have shown a good agreement between the forecasted interface profiles and those actually observed in the model. From the results of the experiments it follows that the separation of the interface into two parts, below and above the semipervious layer, decreases as the length of landward seawater intrusion increases. The proposed solution can readily be applied to the case of a phreatic surface under an artificial recharge basin when the aquifer is separated into subaquifers by a sequence of semipervious layers.

AB-91. Shelat, R.N., Mehta, P.R. and Patel, G.G., "Salt Water Intrusion into Porous Media", Journal of the Institution of Engineers (India), Vol. 55, Part Ph. 2, February, 1975, pp. 55-60.

Experimental model studies were performed to investigate some aspects of subsurface salt water intrusion. The problem of salt water intrusion into coastal aquifers is essentially a problem of stratified fluid flow in porous strata. A boundary surface or interface is formed wherever the two fluids (freshwater and seawater) are in contact. The experimental apparatus comprised a Hele-Shaw Model, pumping arrangement, and controlling devices. The model consisted of two pieces of 12 mm thick plate glass spaced about 2 mm apart and connected to a reservoir consisting of 20 cm x 20 cm x 70 cm high perspex sheet at either end. A well defined interface between lighter and heavier fluids was shown and the Ghyben-Herzberg relationship was satisfied. A type of mixing distinct from diffusion occurred at the interface whenever the interface shifted. The wedge front velocity was determined to be inversely proportional to wedge length and also dependent upon the inclination of impervious beds and density of the intruding fluid. The fresh water piezometric surface must be above the mean sea level a distance equal to S minus one times the distance below sea level to the lowest pumping zone which must be protected in order to prevent sea water intrusion into coastal aquifer, where S is the specific gravity of sea water. For an aquifer of finite thickness, the maintenance of the fresh water surface above sea level will result in seaward loss of fresh water. The relationship between the equilibrium wedge length and the rate of seaward flow of fresh water is independent of the distance of the aquifer below sea level. Correction in overdraft conditions by suitable recharge is a feasible method to check the intrusion.

AB-92. Stegner, S.R., "Analog Model Study of Ground Water Flow in the Rehoboth Bay Area, Delaware", Technical Report No. 12, July, 1972, National Technical Information Service, U.S. Department of Commerce, Springfield, Virginia.

The Rehoboth Bay Area, Delaware, depends on ground water for its fresh water supply, but increased pumping demands may threaten to lower the water table and allow salt water intrusion into the wells. A Hele-Shaw (viscous flow) Analog Model was used in an analogy between flow through porous media and viscous flow in a narrow capillary space. The model makes it possible to use two different glycerin solutions, flowing in a one mm space in the laboratory, to make observations and predict interactions between fresh and salty ground water in nature. Results include the amount of fresh water being discharged to the sea and the extent of salt water intrusion under various imposed natural conditions. If Indian River Inlet were to be closed, whether by locks or by filling, and Rehoboth Bay were allowed to rise and seek its own level, the size of the fresh ground water body would increase, and a more favorably positioned salt water interface would be formed. This would increase significantly the amount of usable fresh water, and pumping would less likely produce salt water pollution of the wells.

MODELING: COASTAL SEAWATER INTRUSION--NUMERICAL

AB-93. Burns, A.W., Frimpter, M.H. and Willey, R.E., "Evaluation of Data Availability and Examples of Modeling for Ground Water Management on Cape Cod, Massachusetts", USGS/WRD-76/003, December, 1975, U.S. Geological Survey, Boston, Massachusetts.

Ground water is the major source of water for public and private supplies on Cape Cod, Massachusetts. A peninsula in the Atlantic Ocean, Cape Cod is underlain by unconsolidated earth materials that contain a lenslike reservoir of fresh water "floating" on salty ground water. Areal and cross sectional computer simulation models based on solutions to the ground water flow equation demonstrate the interdependence of water table altitude, recharge, withdrawal, spatial variations and anisotropy of hydraulic conductivity, spatial variations of specific yield, position of the fresh water/salt water zone of diffusion, and other boundaries. Currently available (1974) descriptions of these parameters, however, are insufficient to calibrate the models to adequately predict future hydrologic conditions for resource planning and management.

AB-94.   Cantatore, W.P. and Volker, R.E., "Numerical Solution of the Steady Interface in a Confined Coastal Aquifer", Institute of Engineers--Australian Civil Engineering Transactions, Vol. CE 16, No. 2, 1974, pp. 115-119.

In the work described, limitations of the Dupuit flow, regular geometry, homogeneity and isotropy of the medium have been overcome by use of the finite element method for solution of the governing differential equations. Solutions obtained included the steady interface, the steady interface subject to pumping from a well in the aquifer, and the steady interface in an anisotropic medium. The results of the steady interface in a confined, homogeneous, isotropic aquifer with and without a pumping well were verified experimentally using a Hele-Shaw Model.

AB-95.   Cheng, R.T., "Finite Element Modeling of Flow Through Porous Media", OWRT-C-4026(9006)(4), March, 1975, Office of Water Research and Technology, U.S. Department of the Interior, Washington, D.C.

Findings of an investigation into the application of the finite element method to ascertain ground water flow problems are presented followed by reproductions of five articles from the published literature originating from the study and an appendix containing a computer subroutine package for solving the convective-dispersion material balance equation using quadratic isoparametric finite elements. The published articles deal with the development of the basic theory for the finite element method as applied to ground water flow as well as basic schemes for application of the method. The results of this development are applied to the problem of sea water encroachment in coastal aquifers, brine movement through causeways, and dispersion of pollutants in ground water flows.

AB-96.   Christensen, B.A., "Mathematical Methods for Analysis of Fresh Water Lenses in the Coastal Zones of the Floridan Aquifer", OWRT-A-032-FLA(1), December, 1978, Office of Water Research and Technology, U.S. Department of the Interior, Washington, D.C.

Concentrated land development in Florida's coastal regions often leads to saltwater intrusion in coastal aquifers. It has been proposed that excess freshwater could be artificially recharged to these coastal aquifers in order to drive out the intruded saltwater by the formation of a freshwater lens. The purpose of this study was to develop methods for analyzing this process. A numerical model based on the finite element method is developed by which the boundaries of a growing freshwater lens can be traced in two horizontal dimensions. This model simulates the injection of freshwater through multiple wells into a thin confined or leaky saline aquifer in which these may be a pre-existing head gradient. The use of the model is demonstrated by application to the peninsula of Pinellas County, Florida. Field data was collected, interpreted and used in a numerical model which simulates interaction between the dredged canals and the water table in the reclaimed land. Finally, a numerical scheme using the method of characteristics is developed to predict the movement in the vertical plane of the interface at the boundary of a freshwater lens in a saline aquifer.

AB-97. Contractor, D.N., "A Two-Dimensional, Finite Element Model of Salt Water Intrusion in Groundwater Systems", Technical Report No. 26, October, 1981, Water Research Institute of the Western Pacific, Guam University, Agana, Guam.

Careful management of aquifers with coastal boundaries is essential to guard against excessive salinity concentrations. Management includes knowing the location of the salt water under present and future pumping conditions. A numerical model can help estimate the location of the salt water for given sets of hydraulic conditions. This report describes the development and use of a sharp interface model to simulate salt water intrusion into coastal and insular ground water aquifers. The model, using linear triangular elements, can simulate steady or unsteady conditions in confined or unconfined aquifers. Three types of boundary conditions can be specified: head as a function of time; flow as a function of time; and a mixed boundary condition for freshwater flowing into a coastal boundary. Freshwater wells can be located at any node. Freshwater and saltwater toes can be tracked with time, and the direction and velocity in each element can be identified. In addition, aquifers containing only freshwater can be simulated by specifying the saltwater head at all nodes to be much less than expected freshwater heads. The model has been used to simulate several natural aquifers, with complex boundaries, and varied basement topography and boundary conditions. The results were compared with analytic solutions and, in all cases, the comparisons were good. The availability of such a model enables the manager of an aquifer to explore the advantages and disadvantages of various options relating to the development of the resource. The computer program has been written in Fortran.

AB-98. Contractor, D.N., "A One-Dimensional, Finite Element Salt Water Intrusion Model", Technical Report No. 20, February, 1981, Water Resources Research Center, Guam University, Agana, Guam.

Ground water resources near the coastline and on islands should be managed in such a way that salt water intrusion into an aquifer does not jeopardize the quality of the water resource. Management of the salt water intrusion includes knowing the location of the wedge and predicting its response to changes in aquifer recharge and pumpage so that its future location can be controlled. A model is needed to make these predictions. The model may be analytical in nature for simple geometries, or numerical for more complicated aquifers. This report describes a one-dimensional, finite element salt water intrusion model, which can be very useful in understanding local features of an extensive aquifer system. A one-dimensional model can be an appropriate tool when first-cut results are necessary in a limited time. Two-dimensional models can be used to provide one-dimensional results, but only by using twice the number of nodes used in a one-dimensional model. Two-dimensional models should be used whenever the necessary data, computer facilities, and budget are available. A one-dimensional model's advantages are its simplicity, low cost of data aquisition, low computer time, and low total cost, as well as its versatility. Confined and unconfined aquifers can be handled, and time-dependent situations can be taken into account. In addition, nonhomogeneities can be specified in the longitudinal direction. The computer program was written in Fortran.

AB-99. Hall, P.L., "Computer Model Guides Salt Water Well Construction", Johnson Drillers Journal, Vol. 47, No. 1, January-February, 1975, pp. 1-2.

A computer model was designed to simulate salt water intrusion and the effects of both fresh and salt water wells on the movement of the salt water wedge. The study was conducted by the Provincial Municipal Engineering Division in New Brunswick, Canada, to avert a serious shortage of fresh water supplies to the fish-processing industry in the area. The computer model showed that it was practical to construct salt water wells that would satisfy these conditions: (1) the salt water wells should not cause contamination of existing wells, (2) the salt water wells should control the extent of salt water intrusion, and (3) the salt water wells should decrease the flow of fresh ground water to the seas and thus provide a fresh water supply for a considerable period of time. The wells were placed in production before the end of 1973 and the salt water movement has been monitored and compared with the predicted movement. Results of the computer model have been of considerable value.

AB-100. Land, L.F., "Effect of Lowering Interior Canal Stages on Salt-Water Intrusion into the Shallow Aquifer in Southeast Palm Beach County, Florida", Open File Report FL 75-74, 1975, U.S. Geological Survey, Tallahassee, Florida.

Land in southeast Palm Beach County, Florida, is undergoing a large scale change in use, from agricultural to residential. To accommodate residential use, a proposal was made to lower canal stages in the interior part of the area undergoing change. The two main tools used in the investigation were a digital model for aquifer evaluation and an analytical technique for predicting the movement of the salt water front in response to a change of ground water flow into the ocean. Test results show that lowering part of the interior canal water levels three feet does not affect the aquifer head or salt water intrusion along the coastal area of Lake Worth. Lowering interior canal water levels by as much as four feet would result in some salt water intrusion.

AB-101. Lee, C. and Cheng, R. T., "On Seawater Encroachment in Coastal Aquifers", Water Resources Research, Vol. 10, No. 5, October, 1974, pp. 1039-1043.

The seawater encroachment in a coastal aquifer is studied by means of a mathematical model. In some aquifers the dispersion amalgamates the encroaching seawater and the discharging freshwater to produce an extensive zone of diffusion. The coupled nonlinear conservation equations of mass and of salt are formulated and solved by the finite element method with the aids of iteration and underrelaxation. The numerical procedure is verified by comparing solutions with known results in the literature. In this study, the seawater encroachment in the Biscayne Aquifer at Cutler area, Florida, is modeled; the numerical results are in good qualitative agreement with field data.

AB-102. Kashef, A.I., "Diffusion and Dispersion in Porous Media--Salt Water Mounds in Coastal Aquifers", Report No. 11, 1968, Water Resources Research Center, North Carolina State University, Raleigh, North Carolina.

A mathematical model of the movement of fresh and salt waters in coastal aquifers in which fresh water is withdrawn by wells is presented. The location of the fresh-salt water interface in artesian and unconfined aquifers is determined on the basis of analysis of hydraulic forces in the body of fresh water. The three-dimensional flow systems are developed by analogy to two-dimensional finite-length systems. Dupuit-Forchheimer assumptions are entirely eliminated. Results compare favorably with the best previous techniques but the method is simpler. A Fortran program for the IBM 360 was used to obtain dimensionless solutions for 130 cases and the results are summarized in tabular and graphic forms. In field cases many results can be attained easily on a desk calculator. A Hele-Shaw Model was constructed to study two-dimensional flow of two immiscible fluids of different viscosities in transient and steady flow. Salt water intrusion was studied combining the effects of natural flow and multiple well pumpage.

AB-103. Kashef, A.I., "Management of Retardation of Salt Water Intrusion in Coastal Aquifers", OWRT-C-4061(9013)(2), July, 1975, Office of Water Research and Technology, U.S. Department of the Interior, Washington, D.C.

The free surfaces in rectangular earth sections or cores of earth dams were determined by the finite element method using a new technique of horizontally displaced mesh. Ten cases were solved and compared with a previously developed and simpler method in which the hydraulic forces were analyzed and which resulted in closed form mathematical solutions. The comparison indicated that the method of hydraulic forces can satisfactorily be used in artesian aquifers to locate the saline water-freshwater interface which is analogous to the free surface in earth dams. The solved cases correspond to a wide range of coverage of practical conditions. The effects of recharge wells under the transient flow conditions were then superimposed on these due to the natural flow in order to determine the degree of salt water retardation. Single wells as well as a battery of recharge wells parallel to the coast line were analyzed. In all cases, dimensionless parameters were used to allow for the changes in the pump capacities and the time intervals. Various well locations were examined and in the case of well batteries various spacings were introduced.

AB-104. Kashef, A.I. and Safar, M.M., "Comparative Study of Fresh-Salt Water Interfaces Using Finite Element and Simple Approaches", Water Resources Bulletin, Vol. 11, No. 4, August, 1975, pp. 651-665.

The fresh-salt water interface in artesian aquifers has been investigated by various techniques on the basis of its analogy to the free surface in earth dams or cores of dams. One of the simple methods was developed by the analysis of hydraulic forces. A suggested finite element method is used for the purpose of comparing with the method of hydraulic forces. The presented procedure eliminates some of the difficulties and uncertainties in current finite element procedures. Both methods proved to be in close agreement. Moreover, the hydraulic heads along the upper boundary of the artesian aquifer were found to be in close agreement with Dupuit's Equation.

AB-105. Kashef, A.I. and Smith, J.C., "Expansion of Salt-Water Zone Due to Well Discharge", Water Resources Bulletin, Vol. 11, No. 6, December, 1975, pp. 1107-1120.

A study has been made to find the extent of the progress of salt-water intrusion due to the operation of one discharge well taking into account various conditions of aquifer properties, pump capacities, natural flows, time effects, and well locations. Dimensionless solutions for specific conditions have been obtained by means of a simple computer program. The range of most common conditions is discussed. One of the main findings of this study was that salt water may be pumped out of a well even if it is located in an initially totally fresh water zone beyond the natural salt/fresh water interface.

AB-106. Pinder, G.F., "Numerical Simulation of Salt-Water Intrusion in Coastal Aquifers, Part 1", OWRT-C-5224(4214)(2), 1975, Office of Water Research and Technology, U.S. Department of the Interior, Washington, D.C.

The set of nonlinear partial differential equations which describe the movement of the salt water front in a coastal aquifer is solved by the Galerkin-finite element method. Pressure and velocities are obtained simultaneously in order to guarantee continuity of velocities between elements. A layered aquifer is modeled either with a functional representation of permeability or by a constant value of permeability over each element. The Galerkin-finite element method is subsequently used to simulate the two-dimensional movement of the salt water front in the Cutler area of the Biscayne Aquifer near Miami, Florida. The concentration distributions for steady state and transient conditions are obtained using pressure head data measured in the field formulating boundary conditions.

AB-107. Sa da Costa, A.A.G. and Wilson, J.L., "A Numerical Model of Seawater Intrusion in Aquifers", NOAA-80021406, December, 1979, National Oceanic and Atmospheric Administration, Rockville, Maryland.

Seawater intrusion along irregular coastlines and under offshore islands is a classical result of ground water development. The development of a numerical model is described: SWIM, a Sea Water Intrusion Model. The finite element method with a Galerkin formulation is used to solve the vertically integrated governing equations (following the "hydraulic" or Dupuit approach). The model accounts for a variety of aquifer situations that occur in the field, such as: homogeneous or nonhomogeneous and isotropic or anisotropic aquifer properties; leaky or nonleaky and phreatic or confined status; and constant or time varying boundary conditions. SWIM simulates situations in which the freshwater either floats over a seawater body, forming a lens, or lies over a finite extent seawater wedge, defining a seawater wedge toe.

AB-108. Segol, G. and Pinder, G.F., "Transient Simulation of Saltwater Intrusion in Southeastern Florida", Water Resources Research, Vol. 12, No. 1, February, 1976, pp. 65-70.

The Galerkin-finite element method can be used to simulate the two-dimensional movement of the saltwater front in the Cutler area of the Biscayne Aquifer near Miami, Florida. The concentration distributions for steady state and transient conditions can be obtained by using pressure head data measured in the field for formulating boundary conditions.

AB-109. Segol, G., Pinder, G.F. and Gray, W.G., "Galerkin-Finite Element Technique for Calculating the Transient Position of the Saltwater Front", Water Resources Research, Vol. 11, No. 2, April, 1975, pp. 343-347.

The set of nonlinear partial differential equations that describe the movement of the saltwater front in a coastal aquifer is solved by the Galerkin-finite element method. Pressure and velocities are obtained simultaneously in order to guarantee continuity of velocities between elements. A layered aquifer is modeled either with a functional representation of permeability or by a constant value of permeability over each element.

AB-110. Segol, G., Pinder, G.F. and Gray, W.G., "Numerical Simulation of Salt-Water Intrusion in Coastal Aquifers, Part 2", OWRT-C-5224(4214)(2),

1975, Office of Water Research and Technology, U.S. Department of the Interior, Washington, D.C.

The computer program developed is intended to solve the two-dimensional equations of ground water flow and mass transport using the Galerkin-finite element method. While the program was developed to investigate saltwater intrusion in coastal aquifers by considering a cross-section of the three-dimensional space, it can also be applied to the study of areal problems. The computer code is written in Fortran IV for the IBM 360/91 system. The method of analysis is briefly described; and related documentation, such as Fortran program listings, are included.

AB-111. Shrivastava, G.S., "Optimisation of Pumping Operations in the El Socorro Aquifer by Mathematical Modelling", <u>Proceedings of the 26th Annual ASCE Hydraulic Division Speciality Conference on Verification of Mathematical and Physical Models in Hydraulic Engineering</u>, August 9-11, 1978, American Society of Civil Engineers, New York, New York, pp. 214-221.

Since 1965 the quality of ground water in the aquifer has deteriorated due to salt water intrusion caused by overpumping and a number of wells have had to be abandoned. In view of the continuing deterioration in water quality it was recognized that unless measures were taken to investigate and control the intrusion the aquifer might become brackish rendering it unsuitable for domestic use. The suitability of various methods for controlling the salt water intrusion was examined and it was found that the development of a monthly pumping schedule, which could so regulate the salt water intrusion as to maintain a recommended chloride concentration in the aquifer, was the only feasible solution. A study was carried out between June, 1972, to July, 1975, in an attempt to develop such a pumping schedule and the purpose of this paper is to briefly describe how this objective was achieved.

AB-112. Tyagi, A.K., "Finite Element Modeling of Sea Water Intrusion in Coastal Ground Water Systems", <u>Proceedings of the 2nd World Congress on Water Resources; Water for Human Needs, New Delhi, India, Dec. 12-16, 1975</u>, Vol. 3, 1975, International Water Resources Association, New Delhi, India, pp. 325-338.

This investigation concerns the hydrodynamics of the intrusion and dispersion of a fresh-salt water interface in coastal ground water systems. Primary attention is focused on developing a model to predict the rate of dispersion of the interface in natural aquifers. Interfacial regimes are classified based on the fresh water flow conditions in an aquifer. Mathematical models including the density and viscosity effects have been mentioned. Dispersion parameters are given and include the properties of porous medium and fluid flow.

AB-113. Ueda, T. et al., "Numerical Analysis of Quasi-three Dimensional Steady Flow and Optimal Well Discharge in Coastal Aquifers",

Transactions of the Japanese Society of Civil Engineers, Vol. 12, November, 1981, pp. 154-156.

This paper discusses the formulations for a numerical analysis of sea water intrusion and for optimal well discharge in consideration of water pollution prevention.

AB-114. Volker, R.E., "Predicting the Movement of Sea Water into a Coastal Aquifer", Technical Paper No. 51, 1980, Australian Water Resources Council, James Cook University of North Queensland, Australia.

Conditions governing the nature of the interface between salt and fresh water when sea water intrudes into a coastal aquifer are outlined. The equations governing the behavior of the interface are formulated for the general case where dispersion occurs, as well as for the case where it is assumed that fresh and salt water are immiscible and an abrupt interface exists between them. For the abrupt interface assumption, solutions are obtained from a Hele-Shaw Experimental Model and from the finite element and boundary integral numerical techniques. The finite element method is applied to confined and unconfined anisotropic aquifers with pumping or recharging near the interface zone for equilibrium conditions. Two alternative formulations are outlined, one using hydraulic head and the other a stream function. Results are obtained for the dispersed interface which show that the extent and location of the mixing zone depend on the dispersion coefficient magnitude and also on the fresh water discharge.

MODELING: NONCOASTAL SALTWATER INTRUSION--PHYSICAL

AB-115. Ahmed, N., "Salt Pollution of Ground Water", OWRR-B-043-MO(1), June 30, 1972, Missouri Water Resources Research Center, University of Missouri at Rolla, Rolla, Missouri.

An experimental model study was conducted to determine the behavior of induced motion in a salt water body overlain by fresh water. For the study, a narrow tank was constructed from plastic sheets of 0.25-inch thickness. The tank's overall length was 100 inches, its depth, 51 inches, and width, 0.06 inch. Salt water of various concentrations was injected always at the same level in the lower part of the tank. Fresh water, added at the top of the tank, was drawn out under the action of a suction head through a 0.25-inch diameter nylon tube fixed at the top of the tank. For a given concentration, the development of the saline mound to a steady state was recorded for various discharges.

AB-116. Jorgensen, D.G., "Geohydrologic Models of the Houston District, Texas", Ground Water, Vol. 19, No. 4, July-August, 1981, pp. 418-428.

The decline of water levels, land subsidence, and salt water intrusion in the Houston District were studied with three different models. Since this complex aquifer system is composed of discontinuous sand layers and easily compressible clay layers, land surface subsidence up to 210 cm at Pasadena and salt water intrusion at a rate of several hundred meters per year have resulted. The first electric analog model in the early 1960's simulated water level declines in the city of Houston, but not at distances from the city. It pointed out that the available data and the two-aquifer conceptual model were not adequate for simulation of ground water systems. There was no attempt to simulate land subsidence. The second analog model in the early 1970's, using new data, simulated water table declines in both aquifers (Chicot and Evangeline) very accurately. The distribution of land subsidence was only approximated. A digital model, developed in 1979 and using five layers, simulated water level declines, water from clay compaction, and land subsidence very accurately, but could not predict salt water intrusion.

AB-117. Kashef, A.I., "Model Studies of Salt Water Intrusion", Water Resources Bulletin, Vol. 6, No. 6, November-December, 1970, pp. 944-967.

Model studies are used to verify theories in ground water flow systems. In complex cases, model studies may be extremely useful, especially when a theoretical rigorous analysis does not exist. The models cannot be considered entirely satisfactory due to the several drawbacks in each type in addition to the normal human errors in experimentation. This paper is concerned with viscous flow models. A brief summary of the other types of models, which may possibly be used in connection with salt water intrusion problems, is also given. Gravity flow systems are analogous to some phases of salt intrusion problems. Problems in oil fields bear general similarities to sea water intrusion zones. In oil fields, gas cycling studies give valuable information to sea water problems.

MODELING: NONCOASTAL SALTWATER INTRUSION--NUMERICAL

AB-118. Chandler, R.L. and McWhorter, D.B., "Upconing of the Salt-Water-Fresh-Water Interface Beneath a Pumping Well", Ground Water, Vol. 13, No. 4, July-August, 1975, pp. 354-359.

The upconing of saline water in response to pumping from an overlying layer of fresh water is investigated by numerical integration of the governing differential equation. The transition zone between the fresh and saline water is idealized as an abrupt interface. Full consideration of the nonlinear boundary conditions on the water table and interface surfaces is included for steady flow toward partially penetrating pumping wells in both isotropic and anisotropic aquifers. There exists an optimum well penetration into the fresh-water layer which permits maximum discharge without salt-water entrainment. The optimum penetration increases as the vertical permeability is reduced relative to the horizontal permeability.

AB-119. Curtis, T.G., Jr., "Simulation of Salt Water Intrusion by Analytic Elements", Dissertation Abstracts International, Vol. 44/11-B, 1983, p. 3486.

A computer model is presented that simulates the movement of an interface separating two incompressible fluids: fresh and salty ground water. The technique is a modification of the method of modeling interface movement by the use of distributed singularities originally presented by De Josselin de Jong. The modifications consist of: (1) formulating the problem in terms of potentials rather than velocities (applicable only for piecewise constant densities); and (2) using complex variable methods to implement boundary conditions. The problem is solved by treating the flow as quasi-steady. A boundary value problem is solved at selected times and the convection of the interface is determined by interpolation of the velocities, which are determined by analytic differentiation of the complex potential function. A Hele-Shaw Model was constructed to verify the shape of the interface and movement with time as predicted by the computer model.

AB-120. Haubold, R.G., "Approximation for Steady Interface Beneath a Well Pumping Fresh Water Overlying Salt Water", Ground Water, Vol. 13, No. 3, May-June, 1975, pp. 254-259.

A mathematical expression was developed which would approximate the steady position and shape of a sharp, upconed interface between fresh and salt water in an aquifer when the fresh water only is being pumped from a well. The computation of the interface shape was based on an empirically derived modification of Muskat's approximation for the height of the cone beneath a well. Differing depths of well penetrations and their effect on the upconed interface were investigated with the approximation. The computed interfaces were compared with corresponding interfaces determined experimentally in a Hele-Shaw model.

AB-121. Kleinecke, D., "Mathematical Modeling of Fresh-Water Aquifers Having Salt Water Bottoms", Report No. 71TMP-47, July, 1971, TEMPO, General Electric Company, Santa Barbara, California.

A method is described for modeling the flow and salinity of water in an aquifer which has a fresh water lens overlying saltwater. Dispersion of salts across the interface is taken into account. The flow model is for general two-dimensional horizontal flow in an aquifer which can be either confined or unconfined at each horizontal location and can change character as the water level changes. Some preliminary calculations are described. A bibliography of the literature on freshwater-saltwater combined flows is included as an appendix.

AB-122. Reddell, D.L. and Sunada, D.K., "Numerical Simulation of Dispersion in Groundwater Aquifers", Hydrology Paper No. 41, June, 1970, Dept. of Agricultural Engineering and Dept. of Civil Engineering, Colorado State University, Fort Collins, Colorado.

A flow equation for a mixture of miscible fluids was derived by combining the law of conservation of mass, Darcy's Law, and an equation of state describing the pressure-volume-temperature-concentration relationship. The result is an equation involving two dependent variables, pressure and concentration. A relationship for determining concentration was derived by expressing a continuity equation for the dispersed tracer. An implicit numerical technique was used to solve the flow equation for pressure, and the method of characteristics with a tensor transformation was used to solve the convective-dispersion equation. The results from the flow equation were used in solving the convective-dispersion equation and the results from the convective-dispersion equation were then used to resolve the flow equation. The computer simulator successfully solved the longitudinal dispersion problem and the longitudinal and lateral dispersion problem. Using the tensor transformation, problems of longitudinal and lateral dispersion were successfully solved in a rotated coordinate system. The computer simulator was used to solve the salt-water intrusion problem. The numerical results for the fresh water head in the aquifer closely matched those obtained analytically.

AB-123. Rofail, N., "Mathematical Model of Stratified Groundwater Flow", Hydrological Science Bulletin, Vol. 22, No. 4, December, 1977, pp. 503-512.

The problem of stratified ground water flow (salt water intrusion) is considered in the general case where the hydrological boundaries can have any shape and the impervious bed can have any configuration. The recharge and discharge can also be considered according to any given function of time. The fluid density is assumed to be constant throughout each layer. The effect at the boundaries is considered, whether these are a river or a continuation of the aquifer with different parameters. The systems of equations are taken in alternative matrix forms. The three time level implicit scheme is used and the multi sweep method is applied for the computation. The numerical procedure of the mathematical model is linearly unconditionally stable.

AB-124. Wang, J.D., "Finite Element Model of 2-D Stratified Flow", Journal of the Hydraulics Division, American Society of Civil Engineers, Vol. 105, No. 12, December, 1979, pp. 1473-1485.

A finite element model for the study of hydrodynamics and salt transport in partially-mixed estuaries is presented. The model is based on the laterally integrated time varying equations of motion, continuity and conservation of salt. The Boussinesq approximation is applied to the resulting two-dimensional equations and a simple equation of state is used to relate salinity to density. Although a rigorous mathematical proof cannot be given it is heuristically postulated that the correct boundary conditions, defining a well-posed problem, must consist of applied stresses or strain rates in cases with internal (Reynolds) stresses included. Without the internal stresses the pressure or the normal strain rate only can be prescribed. A simple application demonstrates the ability of this model to accurately represent the moving surface. Another example illustrates the significance of ocean boundary conditions on the internal solution.

AB-125. Wilson, J.L. and Sa da Costa, A., "Finite Element Simulation of a Saltwater/Freshwater Interface with Indirect Toe Tracking", Water Resources Research, Vol. 18, No. 4, August, 1982, pp. 1069-1080.

The porous media two-layer flow problem involving fluids of different density separated by an immiscible interface is solved using the Galerkin finite element method in one dimension. Each layer is present only over a portion of the domain. The transition from two layers to one constitutes a moving boundary which must be calculated as part of the solution. An indirect numerical procedure using a fixed grid is proposed to track the boundary. The procedure uses the Gaussian quadrature points, a nonlinear variation of layer saturation across those elements containing a moving boundary and an imaginary "extra thickness" of the absent layer.

APPENDIX B

LIST OF COUNTIES BY STATE
AND AGGREGATED SUB-AREAS

STATE	COUNTY	AGGREGATED SUB-AREA (ASA) NUMBER
Alabama	Autauga	307
	Baldwin	308
	Barbour	306
	Bibb	307
	Blount	308
	Bullock	307
	Butler	307
	Calhoun	307
	Chambers	306
	Cherokee	307
	Chilton	307
	Choctaw	308
	Clarke	308
	Clay	307
	Cleburne	307
	Coffee	307
	Colbert	602
	Conecuh	307
	Coosa	307
	Covington	307
	Crenshaw	307
	Cullman	308
	Dale	307
	Dallas	307
	De Kalb	602
	Elmore	307
	Escambia	307
	Etowah	307
	Fayette	308
	Franklin	602
	Geneva	307
	Greene	308
	Hale	308
	Henry	306
	Houston	306
	Jackson	602
	Jefferson	308
	Lamar	308
	Lauderdale	602
	Lawrence	602
	Lee	306
	Limestone	602
	Lowndes	307
	Macon	307
	Madison	602
	Marengo	308
	Marion	308

STATE	COUNTY	AGGREGATED SUB-AREA (ASA) NUMBER
Alabama (continued)	Marshall	602
	Mobile	308
	Monroe	307
	Montgomery	307
	Perry	307
	Pickens	308
	Pike	307
	Randolph	307
	Russell	306
	St. Clair	307
	Shelby	308
	Sumter	308
	Talladega	307
	Tallapoosa	307
	Tuscaloosa	308
	Walker	308
	Washington	308
	Wilcox	307
	Winston	308
Arizona	Apache	1501
	Cochise	1503
	Coconino	1502
	Gila	1503
	Graham	1503
	Greenlee	1503
	Maricopa	1503
	Mohave	1502
	Navajo	1501
	Pima	1503
	Pinal	1503
	Santa Cruz	1503
	Yavapai	1503
	Yuma	1502
Arkansas	Arkansas	801
	Ashley	802
	Baxter	1101
	Benton	1104
	Boone	1101
	Bradley	802
	Calhoun	802
	Carroll	1101
	Chicot	802
	Clark	802
	Cleburne	1101
	Cleveland	802

STATE	COUNTY	AGGREGATED SUB-AREA (ASA) NUMBER
Arkansas (continued)	Columbia	1107
	Conway	1104
	Clay	801
	Craighead	801
	Crawford	1104
	Crittenden	801
	Cross	801
	Dallas	802
	Desha	802
	Drew	802
	Faulkner	1104
	Franklin	1104
	Fulton	1101
	Garland	802
	Grant	802
	Greene	801
	Hempstead	802
	Hot Spring	802
	Howard	1107
	Independence	1101
	Izard	1101
	Jackson	801
	Jefferson	802
	Johnson	1104
	Lafayette	1107
	Lawrence	1101
	Lee	801
	Lincoln	802
	Little River	1107
	Logan	1104
	Lonoke	801
	Madison	1101
	Marion	1101
	Miller	1107
	Mississippi	801
	Monroe	801
	Montgomery	802
	Nevada	802
	Newton	1101
	Ouachita	301
	Perry	1104
	Phillips	801
	Pike	802
	Poinsett	801
	Polk	1107
	Pope	1104
	Prairie	801

STATE	COUNTY	AGGREGATED SUB-AREA (ASA) NUMBER
Arkansas (continued)	Pulaski	1104
	Randolph	1101
	St. Francis	801
	Saline	1104
	Scott	1104
	Searcy	1101
	Sebastian	1104
	Sevier	1107
	Sharp	1101
	Stone	1101
	Union	802
	Van Buren	1101
	Washington	1104
	White	801
	Woodruff	801
	Yell	1104
California	Alameda	1804
	Alpine	1807
	Amador	1803
	Butte	1802
	Calaveras	1803
	Colusa	1802
	Contra Costa	1804
	Del Norte	1801
	El Dorado	1802
	Fresno	1803
	Glenn	1802
	Humboldt	1801
	Imperial	1806
	Inyo	1807
	Kern	1803
	Kings	1803
	Lake	1802
	Lassen	1802
	Los Angeles	1806
	Madera	1803
	Marin	1804
	Mariposa	1803
	Mendocino	1801
	Merced	1803
	Modoc	1802
	Mono	1807
	Monterey	1805
	Napa	1804
	Nevada	1802
	Orange	1806

STATE	COUNTY	AGGREGATED SUB-AREA (ASA) NUMBER
California (continued)	Placer	1802
	Plumas	1802
	Riverside	1806
	Sacramento	1802
	San Benito	1805
	San Bernardino	1806
	San Diego	1806
	San Francisco	1804
	San Joaquin	1803
	San Luis Obispo	1805
	San Mateo	1804
	Santa Barbara	1805
	Santa Clara	1804
	Santa Cruz	1805
	Shasta	1802
	Sierra	1802
	Siskiyou	1801
	Solano	1804
	Sonoma	1804
	Stanislaus	1803
	Sutter	1802
	Tehama	1802
	Trinity	1801
	Tulare	1803
	Tuolumne	1803
	Ventura	1806
	Yolo	1802
	Yuba	1802
Colorado	Adams	1007
	Alamosa	1301
	Arapahoe	1007
	Archuleta	1403
	Baca	1103
	Bent	1102
	Boulder	1007
	Chaffee	1102
	Cheyenne	1010
	Clear Creek	1007
	Conejos	1301
	Costilla	1301
	Crowley	1102
	Custer	1102
	Delta	1402
	Denver	1007
	Dolores	1402
	Douglas	1007

STATE	COUNTY	AGGREGATED SUB-AREA (ASA) NUMBER
Colorado (continued)	Eagle	1402
	Elbert	1007
	El Paso	1102
	Fremont	1102
	Garfield	1402
	Gilpin	1007
	Grand	1402
	Gunnison	1402
	Hinsdale	1402
	Huertano	1102
	Jackson	1007
	Jefferson	1007
	Kiowa	1102
	Kit Carson	1010
	Lake	1102
	La Plata	1403
	Larimer	1007
	Las Animas	1102
	Lincoln	1102
	Logan	1007
	Mesa	1402
	Mineral	1301
	Moffat	1401
	Montezuma	1403
	Montrose	1402
	Morgan	1007
	Otero	1102
	Ouray	1402
	Park	1007
	Phillips	1010
	Pitkin	1402
	Prowers	1102
	Pueblo	1102
	Rio Blanco	1401
	Rio Grande	1301
	Routt	1401
	Saguache	1301
	San Juan	1403
	San Miguel	1402
	Sedgwick	1007
	Summit	1402
	Teller	1102
	Washington	1007
	Weld	1007
	Yuma	1010
Connecticut	Fairfield	104

STATE	COUNTY	AGGREGATED SUB-AREA (ASA) NUMBER
Connecticut (continued)	Hartford	105
	Litchfield	104
	Middlesex	105
	New Haven	104
	New London	104
	Tolland	104
	Windham	104
Delaware	Kent	203
	New Castle	203
	Sussex	205
District of Columbia	District of Columbia	206
Florida	Alachua	304
	Baker	303
	Bay	307
	Bradford	304
	Brevard	304
	Broward	305
	Calhoun	306
	Charlotte	304
	Citrus	304
	Clay	304
	Collier	305
	Columbia	304
	Dade	305
	De Soto	304
	Dixie	304
	Duval	304
	Escambia	307
	Flagler	304
	Franklin	306
	Gadsden	306
	Gilchrist	304
	Glades	305
	Gulf	306
	Hamilton	304
	Hardee	304
	Hendry	305
	Hernando	304
	Highlands	305
	Hillsborough	304
	Holmes	307
	Indian River	305
	Jackson	306
	Jefferson	306

STATE	COUNTY	AGGREGATED SUB-AREA (ASA) NUMBER
Florida (continued)	Lafayette	304
	Lake	304
	Lee	305
	Leon	306
	Levy	304
	Liberty	306
	Madison	304
	Manatee	304
	Marion	304
	Martin	305
	Monroe	305
	Nassau	303
	Okaloosa	307
	Okeechobee	305
	Orange	304
	Osceola	305
	Palm Beach	305
	Pasco	304
	Pinellas	304
	Polk	305
	Putnam	304
	St. Johns	304
	St. Lucie	305
	Santa Rosa	307
	Sarasota	304
	Seminole	304
	Sumter	304
	Suwannee	304
	Taylor	306
	Union	304
	Volusia	304
	Wakulla	306
	Walton	307
	Washington	307
Georgia	Appling	303
	Atkinson	303
	Bacon	303
	Baker	306
	Baldwin	303
	Banks	303
	Barrow	303
	Bartow	307
	Ben Hill	303
	Berrien	304
	Bibb	303

STATE	COUNTY	AGGREGATED SUB-AREA (ASA) NUMBER
Georgia (continued)	Bleckley	303
	Brantley	303
	Brooks	304
	Bryan	303
	Bulloch	303
	Burke	303
	Butts	303
	Calhoun	306
	Camden	303
	Candler	303
	Carroll	306
	Catoosa	601
	Charlton	303
	Chatham	303
	Chattahoochee	306
	Chattooga	307
	Cherokee	307
	Clarke	303
	Clay	306
	Clayton	306
	Clinch	304
	Cobb	306
	Coffee	303
	Colquitt	304
	Columbia	303
	Cook	304
	Coweta	306
	Crawford	306
	Crisp	306
	Dade	601
	Dawson	307
	Decatur	306
	De Kalb	306
	Dodge	303
	Dooly	306
	Dougherty	306
	Douglas	306
	Early	306
	Echols	304
	Effingham	303
	Elbert	303
	Emanuel	303
	Evans	303
	Fannin	601
	Fayette	306
	Floyd	307
	Forsyth	306

STATE	COUNTY	AGGREGATED SUB-AREA (ASA) NUMBER
Georgia (continued)	Franklin	303
	Fulton	306
	Gilmer	307
	Glascock	303
	Glynn	303
	Gordon	307
	Grady	306
	Greene	303
	Gwinnett	306
	Habersham	306
	Hall	306
	Hancock	303
	Haralson	307
	Harris	306
	Hart	303
	Heard	306
	Henry	303
	Houston	303
	Irwin	304
	Jackson	303
	Jasper	303
	Jeff Davis	303
	Jefferson	303
	Jenkins	303
	Johnson	303
	Jones	303
	Lamar	303
	Lanier	304
	Laurens	303
	Lee	306
	Liberty	303
	Lincoln	303
	Long	303
	Lowndes	304
	Lumpkin	306
	McDuffie	303
	McIntosh	303
	Macon	306
	Madison	303
	Marion	306
	Meriwether	306
	Miller	306
	Mitchell	306
	Monroe	303
	Montgomery	303
	Morgan	303
	Murray	307

STATE	COUNTY	AGGREGATED SUB-AREA (ASA) NUMBER
Georgia (continued)	Muscogee	306
	Newton	303
	Oconee	303
	Oglethorpe	303
	Paulding	307
	Peach	303
	Pickens	307
	Pierce	303
	Pike	306
	Polk	307
	Pulaski	303
	Putnam	303
	Quitman	306
	Rabun	303
	Randolph	306
	Richmond	303
	Rockdale	303
	Schley	306
	Screven	303
	Seminole	306
	Spalding	306
	Stephens	303
	Stewart	306
	Sumter	306
	Talbot	306
	Taliaferro	303
	Tattnall	303
	Taylor	306
	Telfair	303
	Terrell	306
	Thomas	306
	Tift	304
	Toombs	303
	Towns	601
	Treutlen	303
	Troup	306
	Turner	304
	Twiggs	303
	Union	601
	Upson	306
	Walker	601
	Walton	303
	Ware	303
	Warren	303
	Washington	303
	Wayne	303
	Webster	306

STATE	COUNTY	AGGREGATED SUB-AREA (ASA) NUMBER
Georgia (continued)	Wheeler	303
	White	306
	Whitfield	307
	Wilcox	303
	Wilkes	303
	Wilkinson	303
	Worth	306
Idaho	Ada	1703
	Adams	1703
	Bannock	1703
	Bear Lake	1601
	Benewah	1701
	Bingham	1703
	Blaine	1703
	Boise	1703
	Bonner	1701
	Bonneville	1703
	Boundary	1701
	Butte	1703
	Camas	1703
	Canyon	1703
	Caribou	1703
	Cassia	1703
	Clark	1703
	Clearwater	1704
	Custer	1704
	Elmore	1703
	Franklin	1601
	Fremont	1703
	Gem	1703
	Gooding	1703
	Idaho	1704
	Jefferson	1703
	Jerome	1703
	Kootenai	1701
	Latah	1704
	Lemhi	1704
	Lewis	1704
	Lincoln	1703
	Madison	1703
	Minidoka	1703
	Nez Perce	1704
	Oneida	1601
	Owyhee	1703
	Payette	1703
	Power	1703

STATE	COUNTY	AGGREGATED SUB-AREA (ASA) NUMBER
Idaho (continued)	Shoshone	1701
	Teton	1703
	Twin Falls	1703
	Valley	1703
	Washington	1703
Illinois	Adams	704
	Alexander	705
	Bond	705
	Boone	703
	Brown	704
	Bureau	704
	Calhoun	704
	Carroll	703
	Cass	704
	Champaign	506
	Christian	704
	Clark	506
	Clay	506
	Clinton	705
	Coles	506
	Cook	403
	Crawford	506
	Cumberland	506
	De Kalb	703
	De Witt	704
	Douglas	506
	Du Page	403
	Edgar	506
	Edwards	506
	Effingham	506
	Fayette	705
	Ford	704
	Franklin	705
	Fulton	704
	Gallatin	505
	Greene	704
	Grundy	704
	Hamilton	506
	Hancock	704
	Hardin	505
	Henderson	703
	Henry	703
	Iroquois	704
	Jackson	705
	Jasper	506
	Jefferson	705

STATE	COUNTY	AGGREGATED SUB-AREA (ASA) NUMBER
Illinois (continued)	Jersey	704
	Jo Daviess	703
	Johnson	505
	Kane	403
	Kankakee	704
	Kendall	704
	Knox	704
	Lake	403
	La Salle	704
	Lawrence	506
	Lee	703
	Livingston	704
	Logan	704
	McDonough	704
	McHenry	403
	McLean	704
	Macon	704
	Macoupin	704
	Madison	705
	Marion	705
	Marshall	704
	Mason	704
	Massac	505
	Menard	704
	Mercer	703
	Monroe	705
	Montgomery	705
	Morgan	704
	Moultrie	705
	Ogle	703
	Peoria	704
	Perry	705
	Piatt	704
	Pike	704
	Pope	505
	Pulaski	505
	Putnam	704
	Randolph	705
	Richland	506
	Rock Island	703
	St. Clair	705
	Saline	505
	Sangamon	704
	Schuyler	704
	Scott	704
	Shelby	705
	Stark	704

STATE	COUNTY	AGGREGATED SUB-AREA (ASA) NUMBER
Illinois (continued)	Stephenson	703
	Tazewell	704
	Union	705
	Vermilion	506
	Wabash	506
	Warren	703
	Washington	705
	Wayne	506
	White	506
	Whiteside	703
	Will	403
	Williamson	705
	Winnebago	703
	Woodford	704
Indiana	Adams	406
	Allen	406
	Bartholomew	506
	Benton	506
	Blackford	506
	Boone	506
	Brown	506
	Carroll	506
	Cass	506
	Clark	505
	Clay	506
	Clinton	506
	Crawford	505
	Daviess	506
	Dearborn	502
	Decatur	506
	De Kalb	406
	Delaware	506
	Dubois	506
	Elkhart	404
	Fayette	503
	Floyd	505
	Fountain	506
	Franklin	503
	Fulton	506
	Gibson	506
	Grant	506
	Greene	506
	Hamilton	506
	Hancock	506
	Harrison	505
	Hendricks	506

STATE	COUNTY	AGGREGATED SUB-AREA (ASA) NUMBER
Indiana (continued)	Henry	506
	Howard	506
	Huntington	506
	Jackson	506
	Jasper	704
	Jay	506
	Jefferson	505
	Jennings	506
	Johnson	506
	Knox	506
	Kosciusko	506
	Lagrange	404
	Lake	403
	La Porte	403
	Lawrence	506
	Madison	506
	Marion	506
	Marshall	404
	Martin	506
	Miami	506
	Monroe	506
	Montgomery	506
	Morgan	506
	Newton	704
	Noble	404
	Ohio	502
	Orange	506
	Owen	506
	Parke	506
	Perry	505
	Pike	506
	Porter	403
	Posey	506
	Pulaski	506
	Putnam	506
	Randolph	506
	Ripley	506
	Rush	506
	St. Joseph	404
	Scott	506
	Shelby	506
	Spencer	505
	Starke	403
	Steuben	404
	Sullivan	506
	Switzerland	502
	Tippecanoe	506

STATE	COUNTY	AGGREGATED SUB-AREA (ASA) NUMBER
Indiana (continued)	Tipton	506
	Union	503
	Vanderburgh	505
	Vermillion	506
	Vigo	506
	Wabash	506
	Warren	506
	Warrick	505
	Washington	506
	Wayne	503
	Wells	506
	White	506
	Whitley	506
Iowa	Adair	1009
	Adams	1009
	Allamakee	703
	Appanoose	1011
	Audubon	1009
	Benton	703
	Black Hawk	703
	Boone	703
	Bremer	703
	Buchanan	703
	Buena Vista	703
	Butler	703
	Calhoun	703
	Carroll	703
	Cass	1009
	Cedar	703
	Cerro Gordo	703
	Cherokee	1009
	Chickasaw	703
	Clarke	703
	Clay	1009
	Clayton	703
	Clinton	703
	Crawford	1009
	Dallas	703
	Davis	703
	Decatur	1011
	Delaware	703
	Des Moines	703
	Dickinson	1009
	Dubuque	703
	Emmet	703
	Fayette	703

STATE	COUNTY	AGGREGATED SUB-AREA (ASA) NUMBER
Iowa (continued)	Floyd	703
	Franklin	703
	Fremont	1009
	Greene	703
	Grundy	703
	Guthrie	703
	Hamilton	703
	Hancock	703
	Hardin	703
	Harrison	1009
	Henry	703
	Howard	703
	Humboldt	703
	Ida	1009
	Iowa	703
	Jackson	703
	Jasper	703
	Jefferson	703
	Johnson	703
	Jones	703
	Keokuk	703
	Kossuth	703
	Lee	703
	Linn	703
	Louisa	703
	Lucas	703
	Lyon	1006
	Madison	703
	Mahaska	703
	Marion	703
	Marshall	703
	Mills	1009
	Mitchell	703
	Monona	1009
	Monroe	703
	Montgomery	1009
	Muscatine	703
	O'Brien	1009
	Osceola	1009
	Page	1009
	Palo Alto	703
	Plymouth	1009
	Pocahontas	703
	Polk	703
	Pottawattamie	1009
	Poweshiek	703
	Ringgold	1011

STATE	COUNTY	AGGREGATED SUB–AREA (ASA) NUMBER
Iowa (continued)	Sac	703
	Scott	703
	Shelby	1009
	Sioux	1006
	Story	703
	Tama	703
	Taylor	1009
	Union	1011
	Van Buren	703
	Wapello	703
	Warren	703
	Washington	703
	Wayne	1011
	Webster	703
	Winnebago	703
	Winneshiek	703
	Woodbury	1009
	Worth	703
	Wright	703
Kansas	Allen	1104
	Anderson	1011
	Atchison	1009
	Barber	1103
	Barton	1103
	Bourbon	1011
	Brown	1009
	Butler	1103
	Chase	1104
	Chautauqua	1104
	Cherokee	1104
	Cheyenne	1010
	Clark	1103
	Clay	1010
	Cloud	1010
	Coffey	1104
	Comanche	1103
	Cowley	1103
	Crawford	1104
	Decatur	1010
	Dickinson	1010
	Doniphan	1009
	Douglas	1010
	Edwards	1103
	Elk	1104
	Ellis	1010
	Ellsworth	1010

STATE	COUNTY	AGGREGATED SUB-AREA (ASA) NUMBER
Kansas (continued)	Finney	1103
	Ford	1103
	Franklin	1011
	Geary	1010
	Gove	1010
	Graham	1010
	Grant	1103
	Gray	1103
	Greeley	1103
	Greenwood	1104
	Hamilton	1103
	Harper	1103
	Harvey	1103
	Haskell	1103
	Hodgeman	1103
	Jackson	1010
	Jefferson	1010
	Jewell	1010
	Johnson	1011
	Kearny	1103
	Kingman	1103
	Kiowa	1103
	Labette	1104
	Lane	1103
	Leavenworth	1010
	Lincoln	1010
	Linn	1011
	Logan	1010
	Lyon	1104
	McPherson	1103
	Marion	1104
	Marshall	1010
	Meade	1103
	Miami	1011
	Mitchell	1010
	Montgomery	1104
	Morris	1104
	Morton	1103
	Nemaha	1009
	Neosho	1104
	Ness	1103
	Norton	1010
	Osage	1011
	Osborne	1010
	Ottawa	1010
	Pawnee	1103
	Phillips	1010

STATE	COUNTY	AGGREGATED SUB-AREA (ASA) NUMBER
Kansas (continued)	Pottawatomie	1010
	Pratt	1103
	Rawlins	1010
	Reno	1103
	Republic	1010
	Rice	1103
	Riley	1010
	Rooks	1010
	Rush	1103
	Russell	1010
	Saline	1010
	Scott	1103
	Sedgwick	1103
	Seward	1103
	Shawnee	1010
	Sheridan	1010
	Sherman	1010
	Smith	1010
	Stafford	1103
	Stanton	1103
	Stevens	1103
	Sumner	1103
	Thomas	1010
	Trego	1010
	Wabaunsee	1010
	Wallace	1010
	Washington	1010
	Wichita	1103
	Wilson	1104
	Woodson	1104
	Wyandotte	1011
Kentucky	Adair	505
	Allen	505
	Anderson	505
	Ballard	505
	Barren	505
	Bath	505
	Bell	507
	Boone	502
	Bourbon	505
	Boyd	502
	Boyle	505
	Bracken	505
	Breathitt	505
	Breckinridge	505
	Bullitt	505

STATE	COUNTY	AGGREGATED SUB-AREA (ASA) NUMBER
Kentucky (continued)	Butler	505
	Caldwell	505
	Calloway	602
	Campbell	502
	Carlisle	801
	Carroll	505
	Carter	502
	Casey	505
	Christian	507
	Clark	505
	Clay	505
	Clinton	507
	Crittenden	505
	Cumberland	507
	Daviess	505
	Edmonson	505
	Elliott	502
	Estill	505
	Fayette	505
	Fleming	505
	Floyd	502
	Franklin	505
	Fulton	801
	Gallatin	505
	Garrard	505
	Grant	505
	Graves	801
	Grayson	505
	Green	505
	Greenup	502
	Hancock	505
	Hardin	505
	Harlan	507
	Harrison	505
	Hart	505
	Henderson	505
	Henry	505
	Hickman	801
	Hopkins	505
	Jackson	507
	Jefferson	505
	Jessamine	505
	Johnson	502
	Kenton	502
	Knott	505
	Knox	507
	Larue	505

STATE	COUNTY	AGGREGATED SUB-AREA (ASA) NUMBER
Kentucky (continued)	Laurel	507
	Lawrence	502
	Lee	505
	Leslie	505
	Letcher	505
	Lewis	502
	Lincoln	505
	Livingston	507
	Logan	505
	Lyon	507
	McCracken	505
	McCreary	507
	McLean	505
	Madison	505
	Magoffin	505
	Marion	505
	Marshall	602
	Martin	502
	Mason	505
	Meade	505
	Menifee	505
	Mercer	505
	Metcalfe	505
	Monroe	505
	Montgomery	505
	Morgan	505
	Muhlenberg	505
	Nelson	505
	Nicholas	505
	Ohio	505
	Oldham	505
	Owen	505
	Owsley	505
	Pendleton	505
	Perry	505
	Pike	502
	Powell	505
	Pulaski	507
	Robertson	505
	Rockcastle	507
	Rowan	505
	Russell	507
	Scott	505
	Shelby	505
	Simpson	505
	Spencer	505
	Taylor	505

STATE	COUNTY	AGGREGATED SUB-AREA (ASA) NUMBER
Kentucky (continued)	Todd	507
	Trigg	507
	Trimble	505
	Union	505
	Warren	505
	Washington	505
	Wayne	507
	Webster	505
	Whitley	507
	Wolfe	505
	Woodford	505
Louisiana	Acadia	803
	Allen	803
	Ascension	803
	Assumption	803
	Avoyelles	802
	Beauregard	803
	Bienville	1107
	Bossier	1107
	Caddo	1107
	Calcasieu	803
	Caldwell	802
	Cameron	803
	Catahoula	802
	Claiborne	802
	Concordia	802
	De Soto	1201
	East Baton Rouge	803
	East Carroll	802
	East Feliciana	803
	Evangeline	803
	Franklin	802
	Grant	802
	Iberia	803
	Iberville	803
	Jackson	802
	Jefferson	803
	Jefferson Davis	803
	Lafayette	803
	Lafourche	803
	La Salle	802
	Lincoln	802
	Livingston	803
	Madison	802
	Morehouse	802
	Natchitoches	1107

STATE	COUNTY	AGGREGATED SUB-AREA (ASA) NUMBER
Louisiana (continued)	Orleans	803
	Quachita	802
	Plaquemines	803
	Pointe Coupee	803
	Rapides	802
	Red River	1107
	Richland	802
	Sabine	1201
	St. Bernard	803
	St. Charles	803
	St. Helena	803
	St. James	803
	St. John Baptist	803
	St. Landry	803
	St. Martin	803
	St. Mary	803
	St. Tammany	803
	Tangipahoa	803
	Tensas	802
	Terrebonne	803
	Union	802
	Vermilion	803
	Vernon	803
	Washington	309
	Webster	1107
	West Baton Rouge	803
	West Carroll	802
	West Feliciana	803
	Winn	802
Maine	Androscoggin	101
	Aroostook	101
	Cumberland	102
	Franklin	101
	Hancock	101
	Kennebec	101
	Knox	101
	Lincoln	101
	Oxford	101
	Penobscot	101
	Piscataquis	101
	Sagadahoc	101
	Somerset	101
	Waldo	101
	Washington	101
	York	102

STATE	COUNTY	AGGREGATED SUB-AREA (ASA) NUMBER
Maryland	Allegany	206
	Anne Arundel	205
	Baltimore	205
	Calvert	205
	Caroline	205
	Carroll	205
	Cecil	203
	Charles	206
	Dorchester	205
	Frederick	206
	Garrett	206
	Harford	205
	Howard	205
	Kent	205
	Montgomery	206
	Prince George's	206
	Queen Anne's	205
	St. Mary's	206
	Somerset	205
	Talbot	205
	Washington	206
	Wicomico	205
	Worcester	205
Massachusetts	Barnstable	103
	Berkshire	104
	Bristol	103
	Dukes	103
	Essex	103
	Franklin	105
	Hampden	105
	Hampshire	105
	Middlesex	103
	Nantucket	103
	Norfolk	103
	Plymouth	103
	Suffolk	103
	Worcester	103
Michigan	Alcona	405
	Alger	401
	Allegan	404
	Alpena	405
	Antrim	404
	Arenac	405
	Baraga	401
	Barry	404

STATE	COUNTY	AGGREGATED SUB-AREA (ASA) NUMBER
Michigan (continued)	Bay	405
	Benzie	404
	Berrien	404
	Branch	404
	Calhoun	404
	Cass	404
	Charlevoix	404
	Cheboygan	405
	Chippewa	401
	Clare	405
	Clinton	404
	Crawford	405
	Delta	402
	Dickinson	402
	Eaton	404
	Emmet	404
	Genesee	405
	Gladwin	405
	Gogebic	401
	Grand Traverse	404
	Gratiot	405
	Hillsdale	404
	Houghton	401
	Huron	405
	Ingham	404
	Ionia	404
	Iosco	405
	Iron	402
	Isabella	405
	Jackson	404
	Kalamazoo	404
	Kalkaska	404
	Kent	404
	Keweenaw	401
	Lake	404
	Lapeer	405
	Leelanau	404
	Lenawee	406
	Livingston	406
	Luce	401
	Mackinac	404
	Macomb	406
	Manistee	404
	Marquette	401
	Mason	404
	Mecosta	404
	Menominee	402

STATE	COUNTY	AGGREGATED SUB-AREA (ASA) NUMBER
Michigan (continued)	Midland	405
	Missaukee	404
	Monroe	406
	Montcalm	404
	Montmorency	405
	Muskegon	404
	Newaygo	404
	Oakland	406
	Oceana	404
	Ogemaw	405
	Ontonagon	401
	Osceola	404
	Oscoda	405
	Otsego	405
	Ottawa	404
	Presque Isle	405
	Roscommon	404
	Saginaw	405
	St. Clair	406
	St. Joseph	404
	Sanilac	406
	Schoolcraft	404
	Shiawassee	404
	Tuscola	405
	Van Buren	404
	Washtenaw	406
	Wayne	406
	Wexford	404
Minnesota	Aitkin	701
	Anoka	701
	Becker	901
	Beltrami	901
	Benton	701
	Big Stone	701
	Blue Earth	701
	Brown	701
	Carlton	401
	Carver	701
	Cass	701
	Chippewa	701
	Chisago	701
	Clay	901
	Clearwater	901
	Cook	401
	Cottonwood	701
	Crow Wing	701

STATE	COUNTY	AGGREGATED SUB-AREA (ASA) NUMBER
Minnesota (continued)	Dakota	701
	Dodge	702
	Douglas	701
	Faribault	701
	Fillmore	702
	Freeborn	703
	Goodhue	702
	Grant	901
	Hennepin	701
	Houston	702
	Hubbard	701
	Isanti	701
	Itasca	701
	Jackson	701
	Kanabec	701
	Kandiyohi	701
	Kittson	901
	Koochiching	901
	Lac qui Parle	701
	Lake	401
	Lake of the Wood	901
	Le Sueur	701
	Lincoln	701
	Lyon	701
	McLeod	701
	Mahnomen	901
	Marshall	901
	Martin	701
	Meeker	701
	Mille Lacs	701
	Morrison	701
	Mower	703
	Murray	701
	Nicollet	701
	Nobles	1006
	Norman	901
	Olmsted	702
	Otter Tail	901
	Pennington	901
	Pine	701
	Pipestone	1006
	Polk	901
	Pope	701
	Ramsey	701
	Red Lake	901
	Redwood	701
	Renville	701

STATE	COUNTY	AGGREGATED SUB-AREA (ASA) NUMBER
Minnesota (continued)	Rice	702
	Rock	1006
	Roseau	901
	St. Louis	401
	Scott	701
	Sherburne	701
	Sibley	701
	Stearns	701
	Steele	702
	Stevens	701
	Swift	701
	Todd	701
	Traverse	901
	Wabasha	702
	Wadena	701
	Waseca	701
	Washington	701
	Watonwan	701
	Wilkin	901
	Winona	702
	Wright	701
	Yellow Medicine	701
Mississippi	Adams	802
	Alcorn	801
	Amite	803
	Attala	802
	Benton	802
	Bolivar	802
	Calhoun	802
	Carroll	802
	Chickasaw	308
	Choctaw	802
	Claiborne	802
	Clarke	309
	Clay	308
	Coahoma	802
	Copiah	309
	Covington	309
	De Soto	802
	Forrest	309
	Franklin	802
	George	309
	Greene	309
	Grenada	802
	Hancock	309
	Harrison	309

STATE	COUNTY	AGGREGATED SUB-AREA (ASA) NUMBER
Mississippi (continued)	Hinds	309
	Holmes	802
	Humphreys	802
	Issaquena	802
	Itawamba	308
	Jackson	309
	Jasper	309
	Jefferson	802
	Jefferson Davis	309
	Jones	309
	Kemper	308
	Lafayette	802
	Lamar	309
	Lauderdale	309
	Lawrence	309
	Leake	309
	Lee	308
	Leflore	802
	Lincoln	309
	Lowndes	308
	Madison	802
	Marion	309
	Marshall	802
	Monroe	308
	Montgomery	802
	Neshoba	309
	Newton	309
	Noxubee	308
	Oktibbeha	308
	Panola	802
	Pearl River	309
	Perry	309
	Pike	309
	Pontotoc	802
	Prentiss	308
	Quitman	802
	Rankin	309
	Scott	309
	Sharkey	802
	Simpson	309
	Smith	309
	Stone	309
	Sunflower	802
	Tallahatchie	802
	Tate	802
	Tippah	801
	Tishomingo	602

STATE	COUNTY	AGGREGATED SUB-AREA (ASA) NUMBER
Mississippi (continued)	Tunica	802
	Union	802
	Walthall	309
	Warren	802
	Washington	802
	Wayne	309
	Webster	802
	Wilkinson	802
	Winston	309
	Yalobusha	802
	Yazoo	802
Missouri	Adair	1011
	Andrew	1009
	Atchison	1009
	Audrain	704
	Barry	1101
	Barton	1104
	Bates	1011
	Benton	1011
	Bollinger	705
	Boone	1011
	Buchanan	1009
	Butler	1101
	Caldwell	1011
	Callaway	1011
	Camden	1011
	Cape Girardeau	705
	Carroll	1011
	Carter	1101
	Cass	1011
	Cedar	1011
	Chariton	1011
	Christian	1101
	Clark	704
	Clay	1011
	Clinton	1011
	Cole	1011
	Cooper	1011
	Crawford	705
	Dade	1011
	Dallas	1011
	Daviess	1011
	De Kalb	1011
	Dent	705
	Douglas	1101
	Dunklin	801

STATE	COUNTY	AGGREGATED SUB-AREA (ASA) NUMBER
Missouri (continued)	Franklin	705
	Gasconade	1011
	Gentry	1011
	Greene	1011
	Grundy	1011
	Harrison	1011
	Henry	1011
	Hickory	1011
	Holt	1009
	Howard	1011
	Howell	1101
	Iron	801
	Jackson	1011
	Jasper	1104
	Jefferson	705
	Johnson	1011
	Knox	704
	Laclede	1011
	Lafayette	1011
	Lawrence	1104
	Lewis	704
	Lincoln	704
	Linn	1011
	Livingston	1011
	McDonald	1104
	Macon	1011
	Madison	801
	Maries	1011
	Marion	704
	Mercer	1011
	Miller	1011
	Mississippi	801
	Moniteau	1011
	Monroe	704
	Montgomery	705
	Morgan	1011
	New Madrid	801
	Newton	1104
	Nodaway	1009
	Oregon	1101
	Osage	1011
	Ozark	1101
	Pemiscot	801
	Perry	705
	Pettis	1011
	Phelps	1011
	Pike	704

STATE	COUNTY	AGGREGATED SUB-AREA (ASA) NUMBER
Missouri (continued)	Platte	1011
	Polk	1011
	Pulaski	1011
	Putnam	1011
	Ralls	704
	Randolph	1011
	Ray	1011
	Reynolds	1101
	Ripley	1101
	St. Charles	705
	St. Clair	1011
	St. Francois	705
	St. Louis	705
	Ste. Genevieve	705
	Saline	1011
	Schuyler	704
	Scotland	704
	Scott	801
	Shannon	1101
	Shelby	704
	Stoddard	801
	Stone	1101
	Sullivan	1011
	Taney	1101
	Texas	1011
	Vernon	1011
	Warren	705
	Washington	705
	Wayne	801
	Webster	1011
	Worth	1011
	Wright	1011
Montana	Beaverhead	1002
	Big Horn	1004
	Blaine	1001
	Broadwater	1002
	Carbon	1004
	Carter	1005
	Cascade	1002
	Chouteau	1002
	Custer	1004
	Daniels	1001
	Dawson	1004
	Deer Lodge	1701
	Fallon	1004
	Fergus	1003

STATE	COUNTY	AGGREGATED SUB-AREA (ASA) NUMBER
Montana (continued)	Flathead	1701
	Gallatin	1002
	Garfield	1003
	Glacier	1002
	Golden Valley	1003
	Granite	1701
	Hill	1001
	Jefferson	1002
	Judith Basin	1003
	Lake	1701
	Lewis and Clark	1002
	Liberty	1002
	Lincoln	1701
	McCone	1001
	Madison	1002
	Meagher	1002
	Mineral	1701
	Missoula	1701
	Musselshell	1003
	Park	1004
	Petroleum	1003
	Phillips	1001
	Pondera	1002
	Powder River	1004
	Powell	1701
	Prairie	1004
	Ravalli	1701
	Richland	1004
	Roosevelt	1001
	Rosebud	1004
	Sanders	1701
	Sheridan	1001
	Silver Bowl	1701
	Stillwater	1004
	Sweet Grass	1004
	Teton	1002
	Toole	1002
	Treasure	1004
	Valley	1001
	Wheatland	1003
	Wibaux	1004
	Yellowstone	1004
	Yellowstone Park	1004
Nebraska	Adams	1010
	Antelope	1008
	Arthur	1008

STATE	COUNTY	AGGREGATED SUB-AREA (ASA) NUMBER
Nebraska (continued)	Banner	1007
	Blaine	1008
	Boone	1008
	Box Butte	1008
	Boyd	1008
	Brown	1008
	Buffalo	1008
	Burt	1008
	Butler	1010
	Cass	1009
	Cedar	1006
	Chase	1010
	Cherry	1008
	Cheyenne	1007
	Clay	1010
	Colfax	1008
	Cuming	1008
	Custer	1008
	Dakota	1009
	Dawes	1008
	Dawson	1008
	Deuel	1007
	Dixon	1006
	Dodge	1008
	Douglas	1009
	Dundy	1010
	Fillmore	1010
	Franklin	1010
	Frontier	1010
	Furnas	1010
	Gage	1010
	Garden	1007
	Garfield	1008
	Gosper	1010
	Grant	1008
	Greeley	1008
	Hall	1008
	Hamilton	1010
	Harlan	1010
	Hayes	1010
	Hitchcock	1010
	Holt	1008
	Hooker	1008
	Howard	1008
	Jefferson	1010
	Johnson	1009
	Kearney	1008

STATE	COUNTY	AGGREGATED SUB-AREA (ASA) NUMBER
Nebraska (continued)	Keith	1008
	Keya Paha	1008
	Kimball	1007
	Knox	1006
	Lancaster	1008
	Lincoln	1008
	Logan	1008
	Loup	1008
	McPherson	1008
	Madison	1008
	Merrick	1008
	Morrill	1007
	Nance	1008
	Nemaha	1009
	Nuckolls	1010
	Otoe	1009
	Pawnee	1009
	Perkins	1010
	Phelps	1008
	Pierce	1008
	Platte	1008
	Polk	1010
	Red Willow	1010
	Richardson	1009
	Rock	1008
	Saline	1010
	Sarpy	1009
	Saunders	1008
	Scotts bluff	1007
	Seward	1010
	Sheridan	1008
	Sherman	1008
	Sioux	1008
	Stanton	1008
	Thayer	1010
	Thomas	1008
	Thurston	1009
	Valley	1008
	Washington	1009
	Wayne	1008
	Webster	1010
	Wheeler	1008
	York	1010
Nevada	Carson City	1604
	Churchill	1604
	Clark	1502

STATE	COUNTY	AGGREGATED SUB-AREA (ASA) NUMBER
Nevada (continued)	Douglas	1604
	Elko	1603
	Esmeralda	1603
	Eureka	1603
	Humboldt	1603
	Lander	1603
	Lincoln	1502
	Lyon	1604
	Mineral	1604
	Nye	1603
	Pershing	1603
	Storey	1604
	Washoe	1604
	White Pine	1603
New Hampshire	Belknap	102
	Carroll	102
	Cheshire	105
	Coos	105
	Grafton	105
	Hillsborough	102
	Merrimack	102
	Rockingham	102
	Strafford	102
	Sullivan	105
New Jersey	Atlantic	203
	Bergen	202
	Burlington	203
	Camden	203
	Cape May	203
	Cumberland	203
	Essex	202
	Gloucester	203
	Hudson	202
	Hunterdon	203
	Mercer	203
	Middlesex	202
	Monmouth	202
	Morris	202
	Ocean	203
	Passaic	202
	Salem	203
	Somerset	202
	Sussex	203
	Union	202
	Warren	203

STATE	COUNTY	AGGREGATED SUB-AREA (ASA) NUMBER
New Mexico	Bernalillo	1302
	Catron	1503
	Chaves	1304
	Colfax	1105
	Curry	1203
	De Baca	1304
	Dona Ana	1302
	Eddy	1304
	Grant	1503
	Guadalupe	1304
	Harding	1105
	Hidalgo	1503
	Lea	1204
	Lincoln	1302
	Los Alamos	1302
	Luna	1302
	McKinley	1501
	Mora	1105
	Otero	1302
	Quay	1105
	Rio Arriba	1302
	Roosevelt	1203
	Sandoval	1302
	San Juan	1403
	San Miguel	1304
	Santa Fe	1302
	Sierra	1302
	Socorro	1302
	Taos	1302
	Torrance	1302
	Union	1105
	Valencia	1302
New York	Albany	201
	Allegany	408
	Bronx	202
	Broome	204
	Cattaraugus	407
	Cayuga	408
	Chautauqua	407
	Chemung	204
	Chenango	204
	Clinton	106
	Columbia	201
	Cortland	204
	Delaware	203
	Dutchess	201

STATE	COUNTY	AGGREGATED SUB-AREA (ASA) NUMBER
New York (continued)	Erie	407
	Essex	106
	Franklin	408
	Fulton	201
	Genesee	408
	Greene	201
	Hamilton	201
	Herkimer	201
	Jefferson	408
	Kings	202
	Lewis	408
	Livingston	408
	Madison	408
	Monroe	408
	Montgomery	201
	Nassau	202
	New York	202
	Niagara	407
	Oneida	201
	Onondaga	408
	Ontario	408
	Orange	201
	Orleans	408
	Oswego	408
	Otsego	204
	Putnam	201
	Queens	202
	Rensselaer	201
	Richmond	202
	Rockland	202
	St. Lawrence	408
	Saratoga	201
	Schenectady	201
	Schoharie	201
	Schuyler	408
	Seneca	408
	Steuben	204
	Suffolk	202
	Sullivan	203
	Tioga	204
	Tompkins	408
	Ulster	201
	Warren	201
	Washington	201
	Wayne	408
	Westchester	202

STATE	COUNTY	AGGREGATED SUB-AREA (ASA) NUMBER
New York (continued)	Wyoming	408
	Yates	408
North Carolina	Alamance	301
	Alexander	302
	Alleghany	504
	Anson	302
	Ashe	504
	Avery	601
	Beaufort	301
	Bertie	301
	Bladen	301
	Brunswick	301
	Buncombe	601
	Burke	302
	Cabarrus	302
	Caldwell	302
	Camden	301
	Carteret	301
	Caswell	301
	Catawba	302
	Chatham	301
	Cherokee	601
	Chowan	301
	Clay	601
	Cleveland	302
	Columbus	302
	Craven	301
	Cumberland	301
	Currituck	301
	Dare	301
	Davidson	302
	Davie	302
	Duplin	301
	Durham	301
	Edgecombe	301
	Forsyth	302
	Franklin	301
	Gaston	302
	Gates	301
	Graham	601
	Granville	301
	Greene	301
	Guilford	302
	Halifax	301
	Harnett	301
	Haywood	601

STATE	COUNTY	AGGREGATED SUB-AREA (ASA) NUMBER
North Carolina (continued)	Henderson	601
	Hertford	301
	Hoke	301
	Hyde	301
	Iredell	302
	Jackson	601
	Johnston	301
	Jones	301
	Lee	301
	Lenoir	301
	Lincoln	302
	McDowell	302
	Macon	601
	Madison	601
	Martin	301
	Mecklenburg	302
	Mitchell	601
	Montgomery	302
	Moore	301
	Nash	301
	New Hanover	301
	Northampton	301
	Onslow	301
	Orange	301
	Pamlico	301
	Pasquotank	301
	Pender	301
	Perquimans	301
	Person	301
	Pitt	301
	Polk	302
	Randolph	302
	Richmond	302
	Robeson	302
	Rockingham	301
	Rowan	302
	Rutherford	302
	Sampson	301
	Scotland	302
	Stanly	302
	Stokes	301
	Surry	302
	Swain	601
	Transylvania	601
	Tyrrell	301
	Union	302
	Vance	301

STATE	COUNTY	AGGREGATED SUB-AREA (ASA) NUMBER
North Carolina (continued)	Wake	301
	Warren	301
	Washington	301
	Watauga	504
	Wayne	301
	Wilkes	302
	Wilson	301
	Yadkin	302
	Yancey	601
North Dakota	Adams	1005
	Barnes	901
	Benson	901
	Billings	1005
	Bottineau	901
	Bowman	1005
	Burke	901
	Burleigh	1005
	Cass	901
	Cavalier	901
	Dickey	1006
	Divide	901
	Dunn	1005
	Eddy	1006
	Emmons	1005
	Foster	1006
	Golden Valley	1005
	Grand Forks	901
	Grant	1005
	Griggs	901
	Hettinger	1005
	Kidder	1005
	La Moure	1006
	Logan	1005
	McHenry	901
	McIntosh	1005
	McKenzie	1005
	McLean	1005
	Mercer	1005
	Morton	1005
	Mountrail	1005
	Nelson	901
	Oliver	1005
	Pembina	901
	Pierce	901
	Ramsey	901
	Ransom	901

STATE	COUNTY	AGGREGATED SUB-AREA (ASA) NUMBER
North Dakota (continued)	Renville	901
	Richland	901
	Rolette	901
	Sargent	901
	Sheridan	1005
	Sioux	1005
	Slope	1005
	Stark	1005
	Steele	901
	Stutsman	1006
	Towner	901
	Traill	901
	Walsh	901
	Ward	901
	Wells	1006
	Williams	1005
Ohio	Adams	502
	Allen	406
	Ashland	503
	Ashtabula	407
	Athens	502
	Auglaize	406
	Belmont	502
	Brown	502
	Butler	503
	Carroll	503
	Champaign	503
	Clark	503
	Clermont	502
	Clinton	502
	Columbiana	502
	Coshocton	503
	Crawford	406
	Cuyahoga	407
	Darke	503
	Defiance	406
	Delaware	503
	Erie	406
	Fairfield	502
	Fayette	503
	Franklin	503
	Fulton	406
	Gallia	502
	Geauga	407
	Greene	503
	Guernsey	503

STATE	COUNTY	AGGREGATED SUB-AREA (ASA) NUMBER
Ohio (continued)	Hamilton	502
	Hancock	406
	Hardin	503
	Harrison	503
	Henry	406
	Highland	503
	Hocking	502
	Holmes	503
	Huron	406
	Jackson	502
	Jefferson	502
	Knox	503
	Lake	407
	Lawrence	502
	Licking	503
	Logan	503
	Lorain	407
	Lucas	406
	Madison	503
	Mahoning	502
	Marion	503
	Medina	407
	Meigs	502
	Mercer	406
	Miami	503
	Monroe	502
	Montgomery	503
	Morgan	503
	Morrow	503
	Muskingum	503
	Noble	502
	Ottawa	406
	Paulding	406
	Perry	503
	Pickaway	503
	Pike	503
	Portage	407
	Preble	503
	Putnam	406
	Richland	503
	Ross	503
	Sandusky	406
	Scioto	503
	Seneca	406
	Shelby	503
	Stark	503
	Summit	407

STATE	COUNTY	AGGREGATED SUB-AREA (ASA) NUMBER
Ohio (continued)	Trumbull	502
	Tuscarawas	503
	Union	503
	Van Wert	406
	Vinton	502
	Warren	502
	Washington	503
	Wayne	503
	Williams	406
	Wood	406
	Wyandot	406
Oklahoma	Adair	1104
	Alfalfa	1103
	Atoka	1107
	Beaver	1105
	Beckham	1106
	Blaine	1105
	Bryan	1107
	Caddo	1106
	Canadian	1105
	Carter	1106
	Cherokee	1104
	Choctaw	1107
	Cimarron	1105
	Cleveland	1105
	Coal	1107
	Comanche	1106
	Cotton	1106
	Craig	1104
	Creek	1104
	Custer	1106
	Delaware	1104
	Dewey	1105
	Ellis	1105
	Garfield	1103
	Garvin	1106
	Grady	1106
	Grant	1103
	Greer	1106
	Harmon	1106
	Harper	1103
	Haskell	1104
	Hughes	1105
	Jackson	1106
	Jefferson	1106
	Johnston	1106

STATE	COUNTY	AGGREGATED SUB-AREA (ASA) NUMBER
Oklahoma (continued)	Kay	1103
	Kingfisher	1103
	Kiowa	1106
	Latimer	1104
	Le Flore	1104
	Lincoln	1105
	Logan	1103
	Love	1106
	McClain	1105
	McCurtain	1107
	McIntosh	1105
	Major	1103
	Marshall	1106
	Mayes	1104
	Murray	1106
	Muskogee	1104
	Noble	1103
	Nowata	1104
	Okfuskee	1105
	Oklahoma	1105
	Okmulgee	1105
	Osage	1104
	Ottawa	1104
	Pawnee	1103
	Payne	1103
	Pittsburg	1105
	Pontotoc	1107
	Pottawatomie	1105
	Pushmataha	1107
	Roger Mills	1106
	Rogers	1104
	Seminole	1105
	Sequoyah	1104
	Stephens	1106
	Texas	1105
	Tillman	1106
	Tulsa	1104
	Wagoner	1104
	Washington	1104
	Washita	1106
	Woods	1103
	Woodward	1105
Oregon	Baker	1703
	Benton	1705
	Clackamas	1705
	Clatsop	1705

STATE	COUNTY	AGGREGATED SUB-AREA (ASA) NUMBER
Oregon (continued)	Columbia	1705
	Coos	1705
	Crook	1702
	Curry	1705
	Deschutes	1702
	Douglas	1705
	Gilliam	1702
	Grant	1702
	Harney	1707
	Hood River	1702
	Jackson	1705
	Jefferson	1702
	Josephine	1705
	Klamath	1801
	Lake	1707
	Lane	1705
	Lincoln	1705
	Linn	1705
	Malheur	1703
	Marion	1705
	Morrow	1702
	Multnomah	1705
	Polk	1705
	Sherman	1702
	Tillamook	1705
	Umatilla	1702
	Union	1704
	Wallowa	1704
	Wasco	1702
	Washington	1705
	Wheeler	1702
	Yamhill	1705
Pennsylvania	Adams	204
	Allegheny	502
	Armstrong	501
	Beaver	502
	Bedford	204
	Berks	203
	Blair	204
	Bradford	204
	Bucks	203
	Butler	502
	Cambria	501
	Cameron	204
	Carbon	203
	Centre	204

STATE	COUNTY	AGGREGATED SUB-AREA (ASA) NUMBER
Pennsylvania (continued)	Chester	203
	Clarion	501
	Clearfield	204
	Clinton	204
	Columbia	204
	Crawford	501
	Cumberland	204
	Dauphin	204
	Delaware	203
	Elk	501
	Erie	407
	Fayette	501
	Forest	501
	Franklin	206
	Fulton	206
	Greene	501
	Huntingdon	204
	Indiana	501
	Jefferson	501
	Juniata	204
	Lackawanna	204
	Lancaster	204
	Lawrence	502
	Lebanon	204
	Lehigh	203
	Luzerne	204
	Lycoming	204
	McKean	501
	Mercer	502
	Mifflin	204
	Monroe	203
	Montgomery	203
	Montour	204
	Northampton	203
	Northumberland	204
	Perry	204
	Philadelphia	203
	Pike	203
	Potter	204
	Schuylkill	204
	Snyder	204
	Somerset	501
	Sullivan	204
	Susquehanna	204
	Tioga	204
	Union	204
	Venango	501

STATE	COUNTY	AGGREGATED SUB-AREA (ASA) NUMBER
Pennsylvania (continued)	Warren	501
	Washington	502
	Wayne	203
	Westmoreland	502
	Wyoming	204
	York	204
Rhode Island	Bristol	103
	Kent	103
	Newport	103
	Providence	103
	Washington	103
South Carolina	Abbeville	303
	Aiken	303
	Allendale	302
	Anderson	303
	Bamberg	302
	Barnwell	302
	Beaufort	302
	Berkeley	302
	Calhoun	302
	Charleston	302
	Cherokee	302
	Chester	302
	Chesterfield	302
	Clarendon	302
	Colleton	302
	Darlington	302
	Dillon	302
	Dorchester	302
	Edgefield	303
	Fairfield	302
	Florence	302
	Georgetown	302
	Greenville	302
	Greenwood	302
	Hampton	302
	Horry	302
	Jasper	302
	Kershaw	302
	Lancaster	302
	Laurens	302
	Lee	302
	Lexington	302
	McCormick	303
	Marion	302

STATE	COUNTY	AGGREGATED SUB-AREA (ASA) NUMBER
South Carolina (continued)	Marlboro	302
	Newberry	302
	Oconee	303
	Orangeburg	302
	Pickens	302
	Richland	302
	Saluda	302
	Spartanburg	302
	Sumter	302
	Union	302
	Williamsburg	302
	York	302
South Dakota	Aurora	1006
	Beadle	1006
	Bennett	1005
	Bon Homme	1006
	Brookings	1006
	Brown	1006
	Brule	1005
	Buffalo	1005
	Butte	1005
	Campbell	1005
	Charles Mix	1005
	Clark	1006
	Clay	1006
	Codington	1006
	Corson	1005
	Custer	1005
	Davison	1006
	Day	1006
	Deuel	1006
	Dewey	1005
	Douglas	1006
	Edmunds	1006
	Fall River	1005
	Faulk	1006
	Grant	701
	Gregory	1005
	Haakon	1005
	Hamlin	1006
	Hand	1006
	Hanson	1006
	Harding	1005
	Hughes	1005
	Hutchinson	1006
	Hyde	1005

STATE	COUNTY	AGGREGATED SUB-AREA (ASA) NUMBER
South Dakota (continued)	Jackson	1005
	Jerauld	1006
	Jones	1005
	Kingsbury	1006
	Lake	1006
	Lawrence	1005
	Lincoln	1006
	Lyman	1005
	McCook	1006
	McPherson	1005
	Marshall	1006
	Meade	1005
	Mcllette	1005
	Miner	1006
	Minnehaha	1006
	Moody	1006
	Pennington	1005
	Perkins	1005
	Potter	1005
	Roberts	701
	Sanborn	1006
	Shannon	1005
	Spink	1006
	Stanley	1005
	Sully	1005
	Todd	1005
	Tripp	1005
	Turner	1006
	Union	1006
	Walworth	1005
	Washabaugh	1005
	Yankton	1006
	Ziebach	1005
Tennessee	Anderson	601
	Bedford	602
	Benton	602
	Bledsoe	601
	Blount	601
	Bradley	601
	Campbell	601
	Cannon	507
	Carroll	801
	Carter	601
	Cheatham	507
	Chester	801
	Claiborne	601

STATE	COUNTY	AGGREGATED SUB-AREA (ASA) NUMBER
Tennessee (continued)	Clay	507
	Cocke	601
	Coffee	602
	Crockett	801
	Cumberland	601
	Davidson	507
	Decatur	602
	De Kalb	507
	Dickson	507
	Dyer	801
	Fayette	801
	Fentress	507
	Franklin	602
	Gibson	801
	Giles	602
	Grainger	601
	Greene	601
	Grundy	601
	Hamblen	601
	Hamilton	601
	Hancock	601
	Hardeman	801
	Hardin	602
	Hawkins	601
	Haywood	801
	Henderson	602
	Henry	602
	Hickman	602
	Houston	507
	Humphreys	602
	Jackson	507
	Jefferson	601
	Johnson	601
	Knox	601
	Lake	801
	Lauderdale	801
	Lawrence	602
	Lewis	602
	Lincoln	602
	Loudon	601
	McMinn	601
	McNairy	801
	Macon	507
	Madison	801
	Marion	601
	Marshall	602
	Maury	602

STATE	COUNTY	AGGREGATED SUB-AREA (ASA) NUMBER
Tennessee (continued)	Meigs	601
	Monroe	601
	Montgomery	507
	Moore	602
	Morgan	601
	Obion	801
	Overton	507
	Perry	602
	Pickett	507
	Polk	601
	Putnam	507
	Rhea	601
	Roane	601
	Robertson	507
	Rutherford	507
	Scott	507
	Sequatchie	601
	Sevier	601
	Shelby	801
	Smith	507
	Stewart	507
	Sullivan	601
	Sumner	507
	Tipton	801
	Trousdale	507
	Unicoi	601
	Union	601
	Van Buren	507
	Warren	507
	Washington	601
	Wayne	602
	Weakley	801
	White	507
	Williamson	507
	Wilson	507
Texas	Anderson	1202
	Andrews	1204
	Angelina	1201
	Aransas	1205
	Archer	1106
	Armstrong	1106
	Atascosa	1205
	Austin	1203
	Bailey	1203
	Bandera	1205
	Bastrop	1204

STATE	COUNTY	AGGREGATED SUB-AREA (ASA) NUMBER
Texas (continued)	Baylor	1106
	Bee	1205
	Bell	1203
	Bexar	1205
	Blanco	1204
	Borden	1204
	Bosque	1203
	Bowie	1107
	Brazoria	1202
	Brazos	1203
	Brewster	1303
	Briscoe	1106
	Brooks	1205
	Brown	1204
	Burleson	1203
	Burnet	1203
	Caldwell	1205
	Calhoun	1205
	Callahan	1204
	Cameron	1305
	Camp	1107
	Carson	1105
	Cass	1107
	Castro	1203
	Chambers	1202
	Cherokee	1201
	Childress	1106
	Clay	1106
	Cochran	1204
	Coke	1204
	Coleman	1204
	Collin	1202
	Collingsworth	1106
	Colorado	1204
	Comal	1205
	Comanche	1203
	Concho	1204
	Cooke	1202
	Coryell	1203
	Cottle	1106
	Crane	1303
	Crockett	1303
	Crosby	1203
	Culberson	1303
	Dallam	1105
	Dallas	1202
	Dawson	1204

STATE	COUNTY	AGGREGATED SUB-AREA (ASA) NUMBER
Texas (continued)	Deaf Smith	1106
	Delta	1107
	Denton	1202
	De Witt	1205
	Dickens	1203
	Dimmit	1205
	Donley	1106
	Duval	1205
	Eastland	1203
	Ector	1204
	Edwards	1205
	Ellis	1202
	El Paso	1302
	Erath	1203
	Falls	1203
	Fannin	1107
	Fayette	1204
	Fisher	1203
	Floyd	1203
	Foard	1106
	Fort Bend	1202
	Franklin	1107
	Freestone	1202
	Frio	1205
	Gaines	1204
	Galveston	1202
	Garza	1203
	Gillespie	1204
	Glasscock	1204
	Goliad	1205
	Gonzales	1205
	Gray	1106
	Grayson	1106
	Gregg	1201
	Grimes	1203
	Guadalupe	1205
	Hale	1203
	Hall	1106
	Hamilton	1203
	Hansford	1105
	Hardeman	1106
	Hardin	1201
	Harris	1202
	Harrison	1107
	Hartley	1105
	Haskell	1203
	Hays	1204

STATE	COUNTY	AGGREGATED SUB-AREA (ASA) NUMBER
Texas (continued)	Hemphill	1105
	Henderson	1202
	Hidalgo	1305
	Hill	1203
	Hockley	1203
	Hood	1203
	Hopkins	1107
	Houston	1202
	Howard	1204
	Hudspeth	1302
	Hunt	1201
	Hutchinson	1105
	Irion	1204
	Jack	1202
	Jackson	1205
	Jasper	1201
	Jeff Davis	1303
	Jefferson	1201
	Jim Hogg	1205
	Jim Wells	1205
	Johnson	1202
	Jones	1203
	Karnes	1205
	Kaufman	1202
	Kendall	1205
	Kenedy	1205
	Kent	1203
	Kerr	1205
	Kimble	1204
	King	1106
	Kinney	1305
	Kleberg	1205
	Knox	1106
	Lamar	1107
	Lamb	1203
	Lampasas	1203
	La Salle	1205
	Lavaca	1205
	Lee	1203
	Leon	1202
	Liberty	1202
	Limestone	1203
	Lipscomb	1105
	Live Oak	1205
	Llano	1204
	Loving	1303
	Lubbock	1203

STATE	COUNTY	AGGREGATED SUB-AREA (ASA) NUMBER
Texas (continued)	Lynn	1203
	McCulloch	1204
	McLennan	1203
	McMullen	1205
	Madison	1202
	Marion	1107
	Martin	1204
	Mason	1204
	Matagorda	1204
	Maverick	1305
	Medina	1205
	Menard	1204
	Midland	1204
	Milam	1203
	Mills	1204
	Mitchell	1204
	Montague	1106
	Montgomery	1202
	Moore	1105
	Morris	1107
	Motley	1106
	Nacogdoches	1201
	Navarro	1202
	Newton	1201
	Nolan	1204
	Nueces	1205
	Ochiltree	1105
	Oldham	1105
	Orange	1201
	Palo Pinto	1203
	Panola	1201
	Parker	1203
	Parmer	1203
	Pecos	1303
	Polk	1202
	Potter	1106
	Presidio	1303
	Rains	1201
	Randall	1106
	Reagan	1204
	Real	1205
	Red River	1107
	Reeves	1303
	Refugio	1205
	Roberts	1105
	Robertson	1203
	Rockwall	1202

STATE	COUNTY	AGGREGATED SUB–AREA (ASA) NUMBER
Texas (continued)	Runnels	1204
	Rusk	1201
	Sabine	1201
	San Augustine	1201
	San Jacinto	1202
	San Patricio	1205
	San Saba	1204
	Schleicher	1204
	Scurry	1204
	Shackelford	1203
	Shelby	1201
	Sherman	1105
	Smith	1201
	Somervell	1203
	Starr	1305
	Stephens	1203
	Sterling	1204
	Stonewall	1203
	Sutton	1305
	Swisher	1106
	Tarrant	1202
	Taylor	1203
	Terrell	1303
	Terry	1204
	Throckmorton	1203
	Titus	1107
	Tom Green	1204
	Travis	1204
	Trinity	1202
	Tyler	1201
	Upshur	1107
	Upton	1303
	Uvalde	1205
	Val Verde	1305
	Van Zandt	1201
	Victoria	1205
	Walker	1202
	Waller	1203
	Ward	1303
	Washington	1203
	Webb	1305
	Wharton	1204
	Wheeler	1106
	Wichita	1106
	Wilbarger	1106
	Willacy	1305
	Williamson	1203

STATE	COUNTY	AGGREGATED SUB-AREA (ASA) NUMBER
Texas (continued)	Wilson	1205
	Winkler	1303
	Wise	1202
	Wood	1201
	Yoakum	1204
	Young	1203
	Zapata	1305
	Zavala	1205
Utah	Beaver	1602
	Box Elder	1601
	Cache	1601
	Carbon	1401
	Daggett	1401
	Davis	1601
	Duchesne	1401
	Emery	1401
	Garfield	1403
	Grand	1402
	Iron	1602
	Juab	1602
	Kane	1403
	Millard	1602
	Morgan	1601
	Piute	1602
	Rich	1601
	Salt Lake	1601
	San Juan	1403
	Sanpete	1602
	Sevier	1602
	Summit	1601
	Tooele	1601
	Uintah	1401
	Utah	1601
	Wasatch	1601
	Washington	1502
	Wayne	1403
	Weber	1601
Vermont	Addison	106
	Bennington	201
	Caledonia	105
	Chittenden	106
	Essex	105
	Franklin	106
	Grand Isle	106
	Lamoille	106

STATE	COUNTY	AGGREGATED SUB-AREA (ASA) NUMBER
Vermont (continued)	Orange	105
	Orleans	106
	Rutland	106
	Washington	105
	Windham	105
	Windsor	105
Virginia	Accomack	205
	Albemarle	205
	Alexandria	206
	Alleghany	205
	Amelia	205
	Amherst	205
	Appomattox	205
	Arlington	206
	Augusta	206
	Bath	205
	Bedford	301
	Bland	504
	Botetourt	205
	Bristol	601
	Brunswick	301
	Buchanan	502
	Buckingham	205
	Buena Vista	205
	Campbell	205
	Caroline	205
	Carroll	504
	Charles City	205
	Charlotte	301
	Charlottesville	205
	Chesapeake	205
	Chesterfield	205
	Clarke	206
	Clifton Forge	205
	Colonial Heights	205
	Covington	205
	Craig	205
	Culpeper	205
	Cumberland	205
	Danville	301
	Dickenson	502
	Dinwiddie	301
	Emporia	301
	Essex	205
	Fairfax	206
	Falls Church	206

STATE	COUNTY	AGGREGATED SUB-AREA (ASA) NUMBER
Virginia (continued)	Fauquier	206
	Floyd	504
	Fluvanna	205
	Franklin	301
	Frederick	206
	Fredericksburg	205
	Galax	504
	Giles	504
	Gloucester	205
	Goochland	205
	Grayson	504
	Greene	205
	Greenville	301
	Halifax	301
	Hampton	205
	Hanover	205
	Harrisonburg	206
	Henrico	205
	Henry	301
	Highland	205
	Hopewell	205
	Isle of Wight	205
	James City	205
	King and Queen	205
	King George	206
	King William	205
	Lancaster	205
	Lee	601
	Lexington	205
	Loudoun	206
	Louisa	205
	Lunenburg	301
	Lynchburg	205
	Madison	205
	Manassas	206
	Manassas Park	206
	Martinsville	301
	Mathews	205
	Mecklenburg	301
	Middlesex	205
	Montgomery	504
	Nansemond	205
	Nelson	205
	New Kent	205
	Newport News	205
	Norfolk	205
	Northampton	205

STATE	COUNTY	AGGREGATED SUB-AREA (ASA) NUMBER
Virginia (continued)	Northumberland	205
	Norton	601
	Nottoway	205
	Orange	205
	Page	206
	Patrick	301
	Petersburg	301
	Pittsylvania	301
	Poquoson	205
	Portsmouth	205
	Powhatan	205
	Prince Edward	205
	Prince George	205
	Prince William	206
	Pulaski	504
	Radford	504
	Rapahannock	205
	Richmond	205
	Roanoke	301
	Rockbridge	205
	Rockingham	206
	Russell	601
	Salem	301
	Scott	601
	Shenandoah	206
	Smyth	601
	Southampton	301
	South Boston	301
	Spotsylvania	205
	Stafford	206
	Staunton	206
	Suffolk	205
	Surry	205
	Sussex	301
	Tazewell	601
	Virginia Beach	205
	Warren	206
	Washington	601
	Waynesboro	206
	Westmoreland	206
	Williamsburg	205
	Winchester	206
	Wise	601
	Wythe	504
	York	205
Washington	Adams	1702

STATE	COUNTY	AGGREGATED SUB-AREA (ASA) NUMBER
Washington (continued)	Asotin	1704
	Benton	1702
	Chelan	1702
	Clallam	1706
	Clark	1705
	Columbia	1702
	Cowlitz	1705
	Douglas	1702
	Ferry	1702
	Franklin	1702
	Garfield	1704
	Grant	1702
	Grays Harbor	1705
	Island	1706
	Jefferson	1706
	King	1706
	Kitsap	1706
	Kittitas	1702
	Klickitat	1702
	Lewis	1705
	Lincoln	1702
	Mason	1706
	Okanogan	1702
	Pacific	1705
	Pend Oreille	1701
	Pierce	1706
	San Juan	1706
	Skagit	1706
	Skamania	1705
	Snohomish	1706
	Spokane	1701
	Stevens	1702
	Thurston	1706
	Wahkiakum	1705
	Walla Walla	1702
	Whatcom	1706
	Whitman	1704
	Yakima	1702
West Virginia	Barbour	501
	Berkeley	206
	Boone	504
	Braxton	504
	Brooke	502
	Cabell	502
	Calhoun	502
	Clay	504

STATE	COUNTY	AGGREGATED SUB-AREA (ASA) NUMBER
West Virginia (continued)	Doddridge	502
	Fayette	504
	Gilmer	502
	Grant	206
	Greenbrier	504
	Hampshire	206
	Hancock	502
	Hardy	206
	Harrison	501
	Jackson	502
	Jefferson	206
	Kanawha	504
	Lewis	501
	Lincoln	502
	Logan	502
	McDowell	502
	Marion	501
	Marshall	502
	Mason	504
	Mercer	504
	Mineral	206
	Mingo	502
	Monongalia	501
	Monroe	504
	Morgan	206
	Nicholas	504
	Ohio	502
	Pendleton	206
	Pleasants	502
	Pocahontas	504
	Preston	501
	Putnam	504
	Raleigh	504
	Randolph	501
	Ritchie	502
	Roane	502
	Summers	504
	Taylor	501
	Tucker	501
	Tyler	502
	Upshur	501
	Wayne	502
	Webster	504
	Wetzel	502
	Wirt	502
	Wood	502
	Wyoming	502

STATE	COUNTY	AGGREGATED SUB-AREA (ASA) NUMBER
Wisconsin	Adams	702
	Ashland	401
	Barron	702
	Bayfield	401
	Brown	402
	Buffalo	702
	Burnett	701
	Calumet	402
	Chippewa	702
	Clark	702
	Columbia	702
	Crawford	702
	Dane	703
	Dodge	703
	Door	402
	Douglas	401
	Dunn	702
	Eau Claire	702
	Florence	402
	Fond du Lac	402
	Forest	402
	Grant	703
	Green	703
	Green Lake	402
	Iowa	703
	Iron	401
	Jackson	702
	Jefferson	703
	Juneau	702
	Kenosha	403
	Kewaunee	402
	La Crosse	702
	Lafayette	703
	Langlade	402
	Lincoln	702
	Manitowoc	402
	Marathon	702
	Marinette	402
	Marquette	402
	Menominee	402
	Milwaukee	403
	Monroe	702
	Oconto	402
	Oneida	702
	Outagamie	402
	Ozaukee	403
	Pepin	702

STATE	COUNTY	AGGREGATED SUB–AREA (ASA) NUMBER
Wisconsin (continued)	Pierce	702
	Polk	701
	Portage	702
	Price	702
	Racine	403
	Richland	702
	Rock	703
	Rusk	702
	St. Croix	701
	Sauk	702
	Sawyer	702
	Shawano	402
	Sheboygan	402
	Taylor	702
	Trempealeau	702
	Vernon	702
	Vilas	702
	Walworth	403
	Washburn	701
	Washington	403
	Waukesha	403
	Waupaca	402
	Waushara	402
	Winnebago	402
	Wood	702
Wyoming	Albany	1007
	Big Horn	1004
	Campbell	1004
	Carbon	1007
	Converse	1007
	Crook	1005
	Fremont	1004
	Goshen	1007
	Hot Springs	1004
	Johnson	1004
	Laramie	1007
	Lincoln	1401
	Natrona	1007
	Niobrara	1005
	Park	1004
	Platte	1007
	Sheridan	1004
	Sublette	1401
	Sweetwater	1401
	Teton	1703
	Uinta	1401

STATE	COUNTY	AGGREGATED SUB-AREA (ASA) NUMBER
Wyoming (continued)	Washakie	1004
	Weston	1005